建筑英语

ARCHITECTURE ENGLISH

主编 常乐 孙元元
编者 冯洋 刘艳红

外语教学与研究出版社
FOREIGN LANGUAGE TEACHING AND RESEARCH PRESS
北京 BEIJING

图书在版编目（CIP）数据

建筑英语 / 常乐，孙元元主编 ；冯洋，刘艳红编. -- 北京 ：外语教学与研究出版社，2020.5（2021.1 重印）
ISBN 978-7-5213-1756-5

Ⅰ. ①建… Ⅱ. ①常… ②孙… ③冯… ④刘… Ⅲ. ①建筑－英语－高等学校－教材 Ⅳ. ①TU

中国版本图书馆 CIP 数据核字（2020）第 102325 号

出 版 人　徐建忠
责任编辑　胡春玲
责任校对　邓红红
封面设计　刘 冬
版式设计　涂 俐
出版发行　外语教学与研究出版社
社　　址　北京市西三环北路 19 号（100089）
网　　址　http://www.fltrp.com
印　　刷　北京京师印务有限公司
开　　本　787×1092　1/16
印　　张　13.5
版　　次　2020 年 6 月第 1 版　2021 年 1 月第 2 次印刷
书　　号　ISBN 978-7-5213-1756-5
定　　价　35.90 元

购书咨询：（010）88819926　电子邮箱：club@fltrp.com
外研书店：https://waiyants.tmall.com
凡印刷、装订质量问题，请联系我社印制部
联系电话：（010）61207896　电子邮箱：zhijian@fltrp.com
凡侵权、盗版书籍线索，请联系我社法律事务部
举报电话：（010）88817519　电子邮箱：banquan@fltrp.com
物料号：317560001

前言

本书将建筑学科最基本的问题以专题英语文章的形式呈现出来，让读者在学建筑的同时强化英语，学语言的同时了解建筑，是编者积累多年建筑英语课程的教学经验，深入了解广大学生与读者的需求精心编写而成的。

全书共有10个单元，每单元由围绕同一个主题的三篇阅读文章组成，每篇文章后有相应的练习。练习内容贴近文章主题，旨在帮助学生熟悉该单元主题内容、掌握相关的建筑基础知识及建筑专业词汇，以及用英语进行专业知识的交流与写作。

本书有如下特点：(1) 选材广泛。既包括建筑专业的基础知识，也介绍了著名的建筑家；既介绍了古罗马建筑的风格，也涉猎了现代中国建筑的特点；(2) 专业性强。所选文章内容具有一定的专业学术水平，涉及到建筑专业的最新知识；(3) 信息量大。全书共有30篇文章，可以使学生通过阅读获取较多的专业信息；(4) 难度适中。文章难度适应在校大学生的实际英语水平，能够满足《大学英语课程教学要求》中较高要求的需要；(5) 图文并茂。本书版式新颖，文中涉及的典型建筑也都有相应的图片，便于学生理解与记忆。书后附录的建筑英语专业词汇表以及推荐书目和网站更是为读者提供了复习和拓展学习的机会，对提高学生阅读兴趣，掌握专业词汇，查阅日常工作资料均有裨益。

本书既可作为高等院校建筑类专业的专业英语教材，也可用于其他各专业学生拓展阶段的选修课程；对建筑感兴趣的读者也能从阅读此书中得到乐趣，学到知识。

由于作者水平有限，疏漏与不足之处在所难免。敬请广大专家、同仁以及读者不吝指教。

编　者

2008年8月

Contents

Contents

Brief Introduction to Architecture

Warming Up

1. Work with a partner, make a list of the types of architecture you know.
2. Do you want to become an architect? If so, what kind of qualifications do you need to obtain and what kind of contribution can you make to the society?

TEXT A

a church

What Is Architecture?

Architecture is the art of building in which human requirements and construction materials are related so as to furnish practical use as well as an aesthetic solution, thus differing from the pure utility of engineering construction.

Architecture can be a structure, a residence, a bridge, a church and a group of buildings.

a bridge

a structure

a residence

a group of buildings

Architecture as an Art

As an art, architecture is essentially abstract and nonrepresentational and involves the manipulation of the relationships of spaces, volumes, planes, masses, and voids.

Some buildings are so beautiful or interesting that they become famous artworks.

▲ Taj Mahal

▼ Egyptian Pyramids

▼ St. Patrick's Cathedral

Architects use shape, form, color and other art elements and principles to design buildings. Architects design buildings with different styles. You can tell a lot by looking at the building's style!

Lines make patterns on the roof of an Italian cathedral.

Ovals, circles and other shapes decorate the ceiling of this dome.

▼ This castle in Hungary has many forms.

The materials used to make this house create interesting textures.

Time in Architecture

Time is also an important factor in architecture, since a building is usually comprehended in a succession of experiences rather than all at once. In most architecture there is no one vantage point from which the whole structure can be understood. The use of light and shadow, as well as surface decoration, can greatly enhance a structure.

The analysis of building types provides an insight into past cultures and eras. Behind each of the greater styles lies not a casual trend nor a vogue, but a period of serious and urgent experimentation directed toward answering the needs of a specific way of life. Climate, methods of labor, available materials, and economy all impose their dictates. Each of the greater styles has been aided by the discovery of new construction methods. Once developed, a method survives tenaciously, giving way only when social changes or new building techniques have reduced it. That evolutionary process is exemplified by the history of modern architecture, which developed from the first uses of structural iron and steel in the mid-19th century.

Until the 20th century there were three great developments in architectural construction–the post-and-lintel, or trabeated system; the arch system, either the cohesive type, employing plastic materials hardening into a homogeneous mass, or the thrust type, in which the loads are received and counterbalanced at definite points; and the modern steel-skeleton system.

Architecture of the Ancient World

In Egyptian architecture, to which some of the earliest extant structures to be called architecture (erected by the Egyptians before 3000) belong, the post-and-lintel system was employed exclusively and produced the earliest stone columnar buildings in history. The architecture of Western Asia from the same era employed the same system; however, arched construction was also known and used. The Chaldaeans and Assyrians, dependent upon clay as their chief material, built vaulted roofs of damp mud bricks that adhered to form a solid shell.

The Evolution of Styles in the Christian Era

The Romans and the early Christians also used the wooden truss for roofing the wide spans of their basilica halls. Byzantine architects experimented with new principles and developed the pendentive, used brilliantly in the 6th century for the Church of Hagia Sophia in Constantinople.

The Romanesque architecture of the early Middle Ages was notable for strong, simple, massive forms and vaults executed in cut stone. In Lombard Romanesque (11th century) the Byzantine concentration of vault thrusts was improved by the device of ribs and of piers to support them. In the 13th century Gothic architecture emerged in perfected form, as in the Amiens and Chartres Cathedrals.

The birth of Renaissance architecture (15th century) inaugurated a period of several hundred years in Western architecture during which the multiple and complex buildings of the modern world began to emerge, while at the same time no new and compelling structural conceptions appeared. The complex, highly decorated Baroque style was the chief manifestation of the 17th-century architectural aesthetic. The Georgian style was among architecture's notable 18th-century expressions. The first half of the 19th century was given over to the classic revival and the Gothic revival.

New World, New Architecture

The architects of the later 19th century found themselves in a world being reshaped by science, industry, and speed. A new eclecticism arose, such as the architecture based on the École des Beaux-Arts, and what is commonly called Victorian architecture in Britain and the United States. The needs of a new society pressed them, while steel, reinforced concrete, and electricity were among the many new technical means at their disposal.

After more than a half-century of assimilation and experimentation, modern architecture, often called the International style, produced an astonishing variety of daring and original buildings, often steel substructures sheathed in glass. The Bauhaus was a strong influence on modern architecture. As the line between architecture and engineering became a shadow, 20th-century architecture often approached engineering. More recently, postmodern architecture, which exploits and expands the technical innovations of modernism while often incorporating stylistic elements from other architectural styles or periods, has become an international movement.

VOCABULARY

New Words

aesthetic *a.* 美学的，艺术的
nonrepresentational *a.* (美术) 抽象的
texture *n.* 手感，质感
comprehend *v.* 理解，领会
vantage point (观察某物的) 有利位置
vogue *n.* 流行，时尚
tenaciously *adv.* 难以改变地
adhere *v.* 粘附，附着
inaugurate *v.* 开始，创始
manifestation *n.* 显示，表明
revival *n.* 复兴，重新流行
eclecticism *n.* 折中主义
assimilation *n.* 吸收，同化

Architectural Terms

void *n.* 孔隙
dome *n.* 圆屋顶，穹顶
post and lintel 柱子与横梁
trabeated system 横梁式结构
arch system 拱券体系
cohesive type 粘合性
homogeneous mass 均质体
thrust type 嵌入式
steel-skeleton system 钢骨体系
columnar building 带圆柱的建筑
vaulted roof 拱形屋顶
wooden truss 木质构架
basilica *n.* 长方形廊柱大厅式建筑
Byzantine *a.* 拜占庭式的
pendentive *n.* 方墙四角圆穹顶支承拱
rib *n.* 拱肋
pier *n.* 支墩
substructure *n.* 基础，下层结构
sheathe *v.* 覆盖，套装

Proper Names

Taj Mahal 泰姬陵 (位于印度阿格拉市，国王沙·贾汗在1629年为其妃所建的陵墓)
St. Patrick's Cathedral 圣巴特里克大教堂
Hungary 匈牙利 (欧洲中部国家)
Chaldaean 迦勒底人 (与巴比伦人血缘相近的闪米特人)
Assyrian 亚述人 (古代生活在两河流域上游的民族，建立了亚述帝国)
Church of Hagia Sophia 圣索菲娅大教堂 (在土耳其伊斯坦布尔市，原为拜占庭帝国东正教的宫廷教堂，拜占庭建筑风格的代表作)
Constantinople 君士坦丁堡 (拜占庭帝国首都，现为土耳其西北部港市伊斯坦布尔)
Romanesque architecture 罗马风建筑 (包含古罗马和拜占庭特色的欧洲建筑风格，该风格盛行于11世纪和12世纪，特点为包括厚实的墙、筒拱穹顶及相对不精细的装饰品)
Lombard Romanesque 伦巴弟罗马式
Gothic architecture 哥特式建筑 (12世纪到15世纪流行于西欧的一种建筑风格，特征是有尖角的拱门，肋形拱顶和飞拱)
Amiens 亚眠 (法国北部城市)
Chartres Cathedrals 夏特尔大教堂 (法国)
Renaissance architecture 文艺复兴时期风格的建筑
Baroque style 巴洛克式风格 (约1550到1700年间盛行于欧洲的一种建筑风格，强调拉紧的效果，其特征是有粗的曲线结构、复杂的装饰和无联系部分间的整体平衡)
Georgian style 乔治王朝时期建筑风格
École des Beaux-Arts (兴起于法国的)装饰艺术风格派
Victorian architecture 维多利亚式建筑
International style 国际风格
Bauhaus (school) 包豪斯建筑学派 (德国建筑研究学派，或指其风格)

EXERCISES

I. Match the English expressions with their Chinese equivalents.

1. basilica
2. pendentive
3. substructure
4. vaulted roof
5. wooden truss
6. trabeated system
7. Gothic architecture
8. Byzantine style
9. Bauhaus
10. thrust type

A. 木质构架
B. 拜占庭风格
C. 方墙四角圆穹顶支承拱
D. 拱形屋顶
E. 长方形廊柱大厅式建筑
F. 哥特式建筑
G. 基础；下层结构
H. 嵌入式
I. 包豪斯建筑学派
J. 横梁式结构

II. Decide whether the following statements are True or False.

1. The use of light and shadow, as well as surface decoration, can enhance a structure.
2. The novel architectural style has nothing to do with the social changes and the discovery and development of new building methods.
3. The Chaldaeans and Assyrians initiated the use of vaulted roofs of damp mud bricks.
4. Romanesque and Gothic architecture dominate the Medieval Age.
5. From Renaissance period and on architecture tended not to be of aesthetic value.
6. In the 20th century, the distinction between architecture and engineering is getting clearer.

III. Choose the best answer to each of the following questions.

1. Which of the following is not mentioned in the text as a form of architecture?

   A. A church. B. An avenue. C. A residence. D. A group of buildings.

2. The following are all manipulated by architecture except ______________.

   A. spaces B. voids C. flows D. masses

3. Which of the following civilizations produced the earliest stone columnar buildings in history?

   A. The Chinese civilization. B. The Indian civilization.
   C. The Egyptian civilization. D. The Greek civilization.

4. Who experimented and created a new building structure pendentive?

   A. Christians. B. Romans. C. Germans. D. Byzantinians.

5. Which of the following is a substyle of Baroque style?

   A. Georgian style. B. Gothic style.
   C. Romanesque style. D. Byzantine style.

IV. Oral task

In next class, you'll be asked to give an oral report based on one of the following questions. Work in teams and search the library or Internet for relevant pictures, facts or stories to support your points.

→ 1 Why is architecture called "an art"?

→ 2 What can be achieved by analyzing the building types?

→ 3 What are the main features of modern architecture?

TEXT B

What Do Architects Do?

Architects are at the forefront of designing the built environment that will surround us in the 21st century. As professional experts in the field of building design and construction, architects use their unique creative skills to advise individuals, property owners and developers, community groups, local authorities and commercial organisations on the design and construction of new buildings, the reuse of existing buildings and the spaces which surround them in our towns and cities.

The work of architects influences every aspect of our built environment, from the design of energy efficient buildings to the integration of new buildings in sensitive contexts. Because of their ability to design and their extensive knowledge of construction, architects' skills are in demand in all areas of property, construction and design. Architects' expertise is invaluable when we need to conserve old buildings, redevelop parts of our towns and cities, understand the impact of a development on a local community, manage a construction programme or need advice on the use and maintenance of an existing building.

Architects work closely with other members of the construction industry including engineers, builders, surveyors, local authority planners and building control officers. Much of their time will be spent visiting sites, assessing the feasibility of a project, inspecting building work or managing the construction process. They will also spend time researching old records and drawings, and testing new ideas and construction techniques.

Society looks to architects to define new ways of living and working, to develop innovative ways of using existing buildings and creating new ones. We need architects' understanding of the complex process of design and construction to build socially and ecologically sustainable cities and communities. Architects can be extremely influential as well as being admired for their imagination and creative skills.

An architect draws plans and pictures.

Architects also build 3-D models to show how the building will look and work.

An architect's plans (or blueprints), drawings and models show construction workers how to build it!

Architects use tools like these:

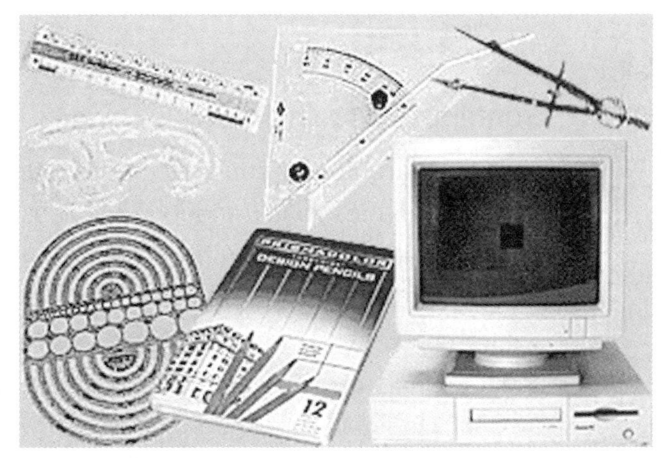

Architects scale, for measuring distances.
Adjustable triangle, for drawing lines and angles.
Compass, for drawing circles and arches.
French curve, for drawing curves.
Circle template, for drawing circles.
Turquoise pencils, for drawing plans.
Computer, for creating plans with design software.

Qualifications

Almost all subjects learnt at school are relevant to architecture, so architectural students should choose the subjects they are strongest in. Although it is not necessary to study art, students should enjoy drawing freehand and making models.

In addition to a good degree, employers are looking for transferable skills–numeracy, interpersonal skills, team working, initiative, decision making and computer literacy. People see these skills as an integral part of architectural education, placing the graduates in a strong position to obtain employment outside the architectural profession in the wider fields of design and business.

Money and Lifestyle

Although it is possible to achieve substantial wealth as architect–and no doubt some architects pursue this as a primary personal goal–it is very improbable. Instead, most architects earn comfortable or modest livings, enjoying reasonable but limited economic stability and prosperity.

Architects begin their careers as wage earners drawing hourly, monthly, or annual salaries which reflect prevailing marketplace conditions. After several years of apprenticeship and further practice, they may become associates or principal owners of firms, either in partnership with others or as sole proprietors. Generally, larger firms provide larger incomes at all levels when contrasted with smaller firms. Thus partners in bigger, well-established offices tend to earn more than partners in firms whose practices are small. Likewise a newly employed draftsperson will probably be paid more by a large firm than by a small one.

Social Status

Social status is an important reason one might choose architecture as a career. An elusive notion at best, it implies the achievement of a certain elevated place in society's hierarchy of who people are and what they do. Social status is relative, meaningful only in comparison with other professions or vocations. Society assumes that architects are educated and that they are both artistically sensitive and technically knowledgeable. Society does not know exactly how architects operate, but it does know that they often create monumental designs for monumental clients. As a result architects may be well respected or admired by members of a social system who, unfortunately, think less of people they consider lacking in education, less talented, and less acceptable in the company of people of wealth, influence, or so-called breeding.

Contributing to Culture

Good architects see themselves as more than professional rendering services to fee-paying clients. Architecture is an expression and embodiment of culture, or cultural conditions. The history of architecture and the history of civilization are inseparable. By designing and building, architects know that they may be contributing directly to culture's inventory of ideas and artifacts, no matter how insignificantly. Thus the search for appropriate cultural achievement is an important motivation for architects.

VOCABULARY

New Words

innovative *a.* 革新的，新颖的
freehand *adv.* 不用绘图仪器地，徒手地
numeracy *n.* 计算能力
initiative *n.* 主动性，创造性
apprenticeship *n.* 学徒身份，学徒年限
proprietor *n.* 所有人，业主
elusive *a.* 难以表述的
elevated *a.* 高贵的，抬高的
hierarchy *n.* 等级制度
embodiment *n.* 具体表现，体现
artifact *n.* 人工制品，艺术品

Architectural Terms

energy efficient building 节能建筑
surveyor *n.* 测量员，检查员
local authority planner 地方权威规划部门
building control officer 建筑管理官员
ecologically sustainable cities and communities 生态可持续型城市和社区
circle template 圆形模板
turquoise pencil 绿松石铅笔

EXERCISES

I. Translate the following English into Chinese and Chinese into English.

1. energy efficient building ____________
2. surveyor ____________
3. ecologically sustainable cities ____________
4. building control officer ____________
5. local authority planner ____________
6. 圆形模板 ____________
7. 等级制度 ____________
8. 学徒身份 ____________
9. 所有人，业主 ____________
10. 人工制品，艺术品 ____________

II. Tell the names of the tools used by architects.

1. ____________
2. ____________
3. ____________
4. ____________
5. ____________
6. ____________
7. ____________

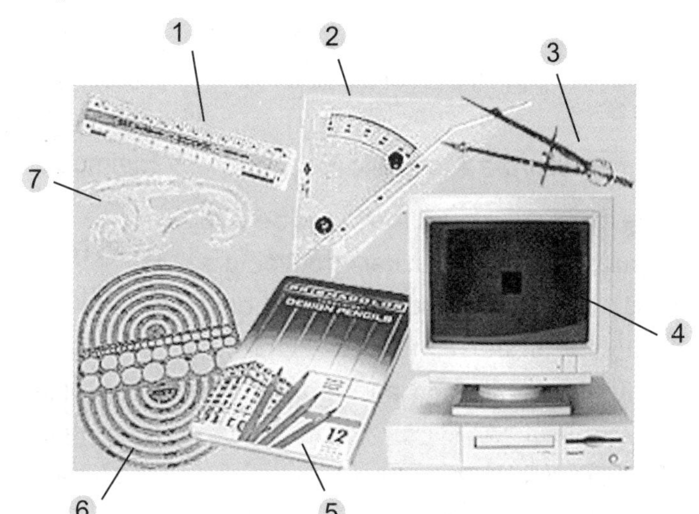

III. Writing task

Suppose you want to choose architecture as a career. Write an application letter to an architecture company for a position of assistant architect. The letter should include:

→ 1 your personal information;

→ 2 the subjects you have learnt in university;

→ 3 your potentials for an assistant architect.

TEXT C

The Pritzker Architecture Prize

The Purpose of the Pritzker Architecture Prize

The purpose of the Pritzker Architecture Prize is to honor annually a living architect whose built work demonstrates a combination of those qualities of talent, vision and commitment, which has produced consistent and significant contributions to humanity and the built environment through the art of architecture.

About the Bronze Medallion

The bronze medallion awarded to each laureate of the Pritzker Architecture Prize is based on the designs of Louis H. Sullivan, famed Chicago architect generally acknowledged as the father of the skyscraper.

On one side is the name of the prize where the laureate's name is also inscribed. On the reverse, the three words inscribed are "firmness", "commodity" and "delight." These are the three conditions referred to by Henry Wotton in his 1624 treatise, *The Elements of Architecture*, which was a translation of thoughts originally set down nearly 2,000 years ago by Marcus Vitruvius in his *Ten Books of Architecture* dedicated to the Roman Emperor Augustus. Wotton who did the translation when he was England's first ambassador to Venice, used the complete quote as: "The end is to build well. Well-building hath three conditions: commodity, firmness and delight."

A Brief History of the Pritzker Architecture Prize

The Pritzker Architecture Prize was established by the Hyatt Foundation in 1979 It has often been described as "architecture' s most prestigious award" or "the Nobel of Architecture".

The prize takes its name from the Pritzker family, whose international business interests are headquartered in Chicago. They have long been known for their support of educational, social welfare, scientific, medical and cultural activities. Jay A. Pritzker, who founded the prize with his wife, Cindy, died on January 23, 1999. His eldest son, Thomas J. Pritzker has become president of the Hyatt Foundation.

In 2004, Chicago celebrated the opening of Millennium Park, in which a music pavilion designed by Pritzker Laureate Frank Gehry was dedicated and named for the founder of the prize. It was in the Jay Pritzker Pavilion that the 2005 awarding ceremony took place.

Tom Pritzker explains, "As native Chicagoans, it's not surprising that our family was keenly aware of architecture, living in the birthplace of the skyscraper, a city filled with buildings designed by architectural legends such as Louis Sullivan, Frank Lloyd Wright, Mies van der Rohe, and many others." He continues, "In 1967, we acquired an unfinished building which was to become the Hyatt Regency Atlanta. Its soaring atrium was wildly successful and became the signature piece of our hotels around the world. It was immediately apparent that this design had a pronounced effect on the mood of our guests and attitude of our employees. While the architecture of Chicago made us cognizant of the art of architecture, our work with designing and building hotels made us aware of the impact architecture could have on human behavior. So in 1978, when we were approached with the idea of honoring living architects, we were responsive. Mom and Dad believed that a meaningful prize would encourage and stimulate not only a greater public awareness of buildings, but also would inspire greater creativity within the architectural profession." He went on to add that he was extremely proud to carry on that effort on behalf of his family.

Procedures

Many of the procedures and rewards of the Pritzker Prize are modeled after the Nobel Prize. Laureates of the Pritzker Architecture Prize receive a $100,000 grant, a formal citation certificate, and since 1987, a bronze medallion. Prior to that year, a limited edition Henry Moore sculpture was presented to each laureate.

Nominations

Nominations are accepted from all nations; from government officials, writers, critics, academicians, fellow architects, architectural societies, or industrialists, virtually anyone who might have an interest in advancing great architecture. The prize is awarded irrespective of nationality, race, creed, or ideology.

Any licensed architect may submit a nomination to the Executive Director for consideration by the jury for the Pritzker Architecture Prize. The nominating procedure is continuous from year to year, closing in November each year. It is sufficient to send an e-mail to the Executive Director with the nominee's name and contact information. Nominations received after the closing are automatically considered in the following calendar year.

There are well over 500 nominees from more than 47 countries to date. The final selection is made by an international jury with all deliberation and voting in secret. The jury normally undertakes deliberations early in the calendar year and the winner is announced in the spring.

The Executive Director may also solicit nominations from past laureates, architects, academics, critics, and professionals involved in cultural endeavors or with special expertise in the field of architecture.

VOCABULARY

New Words

medallion *n.* 大奖章，大勋章
laureate *n.* 获奖者
commodity *n.* 便利，有用
hath (古) =has
prestigious *a.* 有威望的，受尊敬的
dedicate *v.* 为（建筑）举行落成典礼
cognizant *a.* 认识到的
nomination *n.* 提名，推荐
ideology *n.* 思想（体系）
nominee *n.* 被提名者
solicit *v.* 征集

Architectural Terms

skyscraper *n.* 摩天大楼
pavilion *n.* 亭子，阁
atrium *n.* 天井，中庭

Proper Names

the Pritzker Architecture Prize 普利兹克建筑奖
Louis H. Sullivan 路易斯·苏利文（1856–1924，美国建筑师，弗兰克·赖特之师，芝加哥学派的代表人物之一，主张"功能决定形式"，主要作品有芝加哥的会堂大厦、圣路易斯的10层温赖特大厦等）
Henry Wotton 亨利·沃顿（1568–1639，英国外交家及诗人）
Marcus Vitruvius 马库斯·维特鲁威（公元前1世纪，古罗马建筑师，所著《建筑十书》在文艺复兴时期、巴洛克及新古典主义时期成为古典建筑的经典）
Augustus 奥古斯都（63 BC–AD 14，罗马帝国第一代皇帝）
the Hyatt Foundation 海厄特基金会
Jay A. Pritzker 杰伊·普利兹克（芝加哥富商，普利兹克奖的创办者）
Millennium Park 千禧公园（芝加哥）
Frank Gehry 弗兰克·葛瑞（普利兹克建筑奖得主）
Frank Lloyd Wright 弗兰克·劳埃德·赖特（1869–1959，美国著名建筑师，他基于自然形式的特殊建筑风格极大地影响了现代建筑业）
Mies van der Rohe 密斯·凡·德·洛（著名德裔美国建筑师）
Hyatt Regency Atlanta 海厄特行政大厦（亚特兰大市）
Henry Moore 亨利·穆尔（1898–1986，英国雕刻家，按照自然形体和节奏原则而非几何形体作抽象雕刻，代表作有石雕《母与子》等）

Answer the following questions according to the text.

1. What is the purpose of the Pritzker Architecture Prize?
2. What is inscribed on both sides of the medal?
3. Who founded the prize?
4. Who can submit a nomination?
5. How is the final selection made?

Unit 2 Architectural History

Warming Up

1. Can you name some architectural styles popularized in history?
2. List the top three world architecture that impress you most and tell your partner their unique features.

TEXT A

The Brief History of World Architecture (I)

Architecture is the practice of building design and its resulting products; customary usage refers only to those designs and structures that are culturally significant. Architecture must satisfy its intended uses, must be technically sound, and must convey aesthetic meaning. The best buildings are often so well constructed that they outlast their original use, surviving not only as beautiful objects, but as documents of the history of cultures.

For Western readers, the architecture of the ancient world, of the Orient, and of the pre-Columbian Americas may be divided into two groups: indigenous architecture and classical architecture. Indigenous architecture includes ways of building that appear to have developed independently in isolated, local cultural conditions. Classical architecture includes the systems and building methods of Greece and Rome.

Indigenous Architecture

The oldest designed environments stable enough to have left architectural traces date from the first development of cities. The Assyrian city of Khorsabad, built during the reign of Sargon II (722-705 BC) and excavated as early as 1842, became the basis for the study of Mesopotamian architecture. Over a long period, the urban culture of Egypt erected the most awesome of the world's ancient monuments. The largest and best preserved pyramids—used for royal tombs—are those of Khufu (2570? BC) and Khafre (2530? BC) at Giza.

Early Indian stone architecture, which was elaborately carved, was more like sculpture than building. At sites such as Ellora and Ajanta, northeast of Bombay, are great series of artificial caves, sculpted out of the solid rock of cliffs. The Chinese house, built in rectangular and symmetrical fashion, reflects a traditional focus on social order. Japanese house design is more concerned with achieving a satisfying relationship with earth, water, rocks, and trees. This approach is epitomized in the Katsura Detached Palace

(17th century).

The Teotihuacán culture (100 BC-AD 700) of Mexico contained two immense pyramids embedded in what was a vast city. The Mayan civilization's greatest building periods fall within the 4th to the 11th century. The Maya created impressive structures through extensive earth moving. Their bold architectural sculpture is either integral with the stone monuments or is added as stucco ornamentation. The Inca Empire was centered in the Peruvian Andes and flourished from about 1200 to 1533. Inca masonry craftsmanship is unexcelled; enormous pieces of stone were transported over mountainous terrain and fitted together with precision.

Classical Architecture

The building systems and forms of ancient Greece and Rome directly determined the course of Western architecture. Forms of the Greek temple, the archetypal shrine of all time, range from the tiny Temple of Nike Apteros (427-424 BC) of about 6 by 9 meter (about 20 by 30 ft), on the Acropolis in Athens, to the gigantic Temple of Zeus (500? BC) at Agrigento in Sicily, which covered more than one hectare (more than two acres).

Two Greek architectural orders developed more or less concurrently. The Doric order predominated on the mainland and in the western colonies. The acknowledged Doric masterpiece is the Parthenon (447-432 BC), which crowns the Athens Acropolis. The Ionic order originated in the cities on the islands and coasts of Asia Minor. It featured capitals with spiral volutes, slender shafts, and elaborate bases. The Corinthian order, a later development, introduced Ionic capitals detailed with acanthus leaves.

Rome became a powerful, well-organized empire, marked by great engineering works—roads, canals, bridges, and aqueducts. Two Roman inventions allowed for greater architectural flexibility: the dome and the groin vault—formed by the intersection of two identical barrel vaults over a square plan. The Romans also introduced the commemorative or triumphal arch and the colosseum or stadium. Rome was noteworthy for grandiose urban design, demonstrated through the forum, complete with basilica, temple, and other features. The most remarkable forum is Hadrian's Villa (125-132) near Tivoli.

The Architecture of Christianity

In the 4th century, Roman Emperor Constantine the Great converted to Christianity and created a Christian empire, prompting the building of many new churches. The surviving basilicas in Rome that most clearly evoke the Early Christian character include Sant' Agnese Fuori le Mura (rebuilt in 630 and later) and Santa Sabina (422-432). Byzantine churches, domed and decorated with mosaics, proliferated throughout the Byzantine Empire; most notable is the Hagia Sophia (532-537), built by Eastern emperor Justinian I at Constantinople.

The basilican plan was elaborated in Romanesque architecture. Some greatest monastic Romanesque churches did not survive the French Revolution (1789-1799) but has been reconstructed in drawings. Its design influenced Romanesque and Gothic churches in Burgundy and beyond.

At the beginning of the 12th century, Romanesque was transformed into Gothic architecture. Although the change was a response to a growing rationalism in Christian theology, it was also the result of technical developments in vaulting. About 1100, the builders of Durham Cathedral in England invented a new method that gave a new geometric articulation—the ribbed vault.

Other developments were the pointed arch and vault, and the flying buttress, which allowed construction of more elegant, higher, and apparently lighter structures. The major French Gothic cathedrals include those at Paris (see Notre Dame), Chartres, Reims, and Amiens. The English Gothic cathedrals include Canterbury, Lincoln, York Minster, and Exeter.

(To be continued)

VOCABULARY

New Words

pre-Columbian *a.* 哥伦布到达美洲以前的
indigenous *a.* 当地的，本土的
reign *n.* 统治，(君主)统治时期
excavate *v.* 发掘(古物等)
awesome *a.* 令人敬畏的
rectangular *a.* 矩形的，具有矩形形状的
symmetrical *a.* 对称的或呈匀称状的
epitomize *v.* 成为……的典型范例
bold *a.* 粗线条的
integral *a.* 构成整体所需要的
Mayan *a.* 马雅的，马雅人的，马雅文化的
Peruvian *a.* 秘鲁的
flourish *v.* 繁荣，兴旺
unexcelled *a.* 无可比拟的
terrain *n.* 地形，地势
precision *n.* 精确，精确性
archetypal *a.* 典型的，原型的
gigantic *a.* 巨大的，庞大的
concurrently *adv.* 并发地，同时地
predominate *v.* 统治，支配，占优势
elaborate *a.* 复杂精美的
intersection *n.* 交汇，相交
commemorative *a.* 表示纪念的
triumphal *a.* 胜利的，凯旋的
noteworthy *a.* 值得注意的
grandiose *a.* 宏伟的，壮丽的
proliferate *v.* 迅速大量地产生
rationalism *n.* 唯理论，理性主义
geometric *a.* 几何的，几何学的

Architectural Terms

stucco *n.* (涂建筑物外墙用的)灰泥
masonry *n.* 石工技艺，砖石建筑
shrine *n.* 神殿，神龛，圣祠
Doric order 多利斯柱型(纯朴、古老的希腊建筑风格)
Ionic *a.* 爱奥尼亚(式样)的(一种古老的希腊建筑风格)
spiral volute 螺旋涡形，涡形花样
shaft *n.* 柱身
Corinthian order 科林斯式柱型(有带叶形饰的钟状柱顶的柱型，为希腊柱型中最华丽者)
acanthus *n.* 叶形装饰
aqueduct *n.* 高架渠，桥管输水道
groin vault 交叉拱顶，十字拱顶
barrel vault 筒形(桶形)拱顶
triumphal arch (为纪念胜利而建的)凯旋门
colosseum *n.* 古罗马椭圆形剧场
forum *n.* (古罗马城镇)用于公开讨论的广场(或市场)
mosaic *n.* 马赛克图案，马赛克切割和镶嵌
vaulting *n.* 拱顶营造，造拱术
articulation *n.* 接合，连接方法
ribbed vault 肋拱
pointed arch 尖券
pointed vault 尖拱
flying buttress 飞券，飞拱

Proper Names

Khorsabad 霍萨巴德（古亚述帝国的一个城市）
Sargon II 萨尔贡二世（古亚述帝国的君主）
Mesopotamia 美索不达米亚（古代西南亚介于底格里斯河和幼发拉底河之间的一个地区，位于现在的伊拉克境内）
Khufu 胡夫（古埃及第四王朝第二代国王，下令建造最大的金字塔）
Khafre 哈夫拉（古埃及第四王朝国王，于公元前约2530年建哈夫拉金字塔和狮身人面像）
Giza 吉萨（埃及东北部城市，同开罗隔河相望；南郊8公里之利比亚沙漠中有著名的金字塔、狮身人面像等古迹）
Ellora & Ajanta 埃罗喇和阿旃陀（位于印度中西部的古村，附近岩洞可追溯到约公元前200年至公元650，藏有辉煌的佛教艺术典范）
Bombay 孟买（印度西部港市）
Katsura Detached Palace 桂离宫（日本）
Teotihuacán 提奥帝华坎（位于今天墨西哥城东北的中墨西哥的古城，其遗址包括太阳金字塔和克萨尔科多神庙）
Maya 马雅人（中美洲古印第安人的一族，具有高度文明）
Inca 印加人（秘鲁高原上克丘亚部落的一支，被西班牙征服前建立了一个北起厄瓜多尔南到智利中部的帝国）
Andes 安第斯山脉（南美洲西部巨大山脉）
Nike Apteros （希腊神话）胜利女神
the Acropolis 雅典卫城
Athens 雅典（希腊的首都）
Zeus 宙斯（希腊神话中的主神、万神之王，天界的统治者）
Agrigento 阿格里琴托（意大利西西里岛西南的一城市，俯瞰地中海）
Sicily 西西里（意大利南部岛屿，位于意大利半岛南端以西的地中海）
Parthenon 帕台侬神殿（希腊神话中女神雅典娜的主要神庙，位于雅典卫城上，建于公元前447年和公元前432年之间，多利安式建筑的杰出代表）
Asia Minor 小亚细亚（黑海与地中海之间亚洲西部的一个半岛，总体上与土耳其范围相当）
Hadrian 哈德良（古罗马皇帝，117–138年在位）
Tivoli 提沃利（意大利中部城市，位于罗马东北偏东）
Constantine the Great 康斯坦丁大帝（古罗马皇帝，288–337年在位，建立了君士坦丁堡）
Sant' Agnese Fuori le Mura 圣阿格涅斯教堂
Santa Sabina 圣萨比纳教堂
Justinian I 查士丁尼一世（483–565，东罗马帝国皇帝，527–565年在位）
Burgundy 勃艮第（法国中东部地区）
Christian theology 基督教义，基督神学
Durham Cathedral 都汉姆大教堂（英国）
Notre Dame （法语）圣母
Reims 兰斯（法国东北部城市）
Canterbury 坎特伯雷（英格兰东南部城市）
Lincoln 林肯市（英格兰东部城市）
York Minster 约克大教堂（英国）
Exeter 埃克塞特（英格兰西南部城市）

EXERCISES

I. Match the Chinese expressions with their English equivalents.

1. stucco	A. 马赛克图案，马赛克切割和镶嵌
2. acanthus	B. 灰泥
3. spiral volute	C. 雅典卫城
4. triumphal arch	D. 叶形装饰
5. Corinthian order	E. 凯旋门
6. vaulting	F. 螺旋涡形，涡形花样
7. ribbed vault	G. 飞券，飞拱，飞扶壁
8. flying buttress	H. 科林斯柱型
9. mosaic	I. 拱顶营造，造拱术
10. the Acropolis	J. 肋拱

II. Decide whether the following statements are True or False.

1. Architecture must satisfy its intended uses, must be technically sound, and must convey aesthetic meaning.
2. For whatever readers, the architecture of the ancient world, Asia, and the pre-Columbian Americas may be divided into two groups: indigenous architecture and classical architecture.
3. Greek civilization developed altogether two architectural orders—Doric and Ionic.
4. Rome was noteworthy for grandiose urban design, demonstrated through the forum, complete with basilica, temple, and other features.
5. One of the most notable Byzantine buildings is the Hagia Sophia, built by Western Roman Emperor Justinian I at Constantinople.
6. Pointed arch and vault and flying buttress allowed construction of more elegant, higher, and apparently lighter structures.

III. Choose the best answer to each of the following questions.

1. The urban culture of ________ erected the most awesome of the world's ancient monuments including the Great Pyramids.

   A. Mesopotamia　　B. Babylonia　　C. Egypt　　D. India

2. The building systems and forms of ancient _________ directly determined the course of Western architecture.

   A. Egypt and Greece　　B. Egypt and Rome

   C. Greece and Assyria　　D. Greece and Rome

3. Byzantine architecture is characterized as being decorated with massive _______.
   A. paintings B. sculptures C. mosaics D. jewelry
4. The forms of early churches were inspired by _______ survived in Rome.
   A. palace B. basilica C. dome D. groin vault
5. At the beginning of the 12th century, Romanesque was transformed into _______ architecture.
   A. Gothic B. classical C. Byzantine D. Roman

IV. Oral task

In next class, you'll be asked to give an oral report based on one of the following questions. Work in teams and search the library or Internet for relevant pictures, facts or stories to support your points.

→ 1 How do you understand classical architecture?

→ 2 What superiority did Roman architecture have?

→ 3 What is the major form of Gothic architecture and what are the features of it?

TEXT B

The Brief History of World Architecture (II)

The Architecture of Islam

During the 7th century the rise of Islam urged the creation of the mosque. The basic plan of an Islamic mosque is the same today as in ancient times. The oldest mosque in Iraq, now a brick ruin, is at Sāmarrā' (847-852). Notable Spanish mosques include the Great Mosque (786-965) at Córdoba, and the Alhambra (1354-1391) at Granada. Islamic architecture in India includes the Tomb of Humayun (1564-1573) in Delhi and the Taj Mahal (1632-1648) in Āgra.

Islam forbade the representation of people and animals, yet craftsmen created highly ornamented buildings. The motifs—using such materials as glazed tile, marble, and mosaic—are geometrical designs, floral arabesques, and Arabic calligraphy.

Renaissance Architecture

The cultural revolution in Western civilization now called the Renaissance brought a revival of the principles and styles of ancient Greek and Roman architecture. Beginning in Italy in about 1400, it spread to the rest of Europe during the next 150 years. Italian architect Filippo Brunelleschi, with his dome design for the Florence Cathedral (1420-1436, Florence, Italy) stood at the threshold between Gothic and Renaissance.

In the 16th century, Rome became the leading center for the new architecture. Saint Peter's Basilica in Vatican City was the most important of many 16th-century projects. Toward the mid-16th century such leading Italian architects as Michelangelo, Baldassare Peruzzi, Giulio Romano, and Giacomo da Vignola began to use the classical Roman elements in ways that became known as mannerism. Arches, columns, and entablatures came to be used as devices to introduce drama through depth recession, asymmetry, and unexpected proportions and scales.

Baroque and Rococo Architecture

Italians were the pioneers of Baroque; the best known was architect-sculptor Gianlorenzo Bernini, designer of the great oval plaza (begun in 1656) in front of Saint Peter's Basilica. The best 17th-century French talent was absorbed in the service of French king Louis XIV to rebuild the Château de Versailles on a regal scale. In England, architect Sir Christopher Wren rose to prominence. His masterpiece is Saint Paul's Cathedral (1675-1710).

When Louis XIV died in 1715, changes in the artistic climate led to the exuberant Rococo style, which came to full flower in Bavaria and Austria through such buildings as the Austrian Benedictine Abbey (1748-1754) at Ottobeuren. Neoclassicism gripped England, resulting in the Georgian style; the North American version became known as the Federal style.

The Industrial Age

The Industrial Revolution, which began in England about 1760, brought a flood of new building materials—for example, cast iron, steel, and glass. Late 18th-century designers and patrons turned toward the original Greek and Roman prototypes, and selective borrowing from another time and place became fashionable. Doric and Ionic columns, entablatures, and pediments were applied to public buildings and important town houses in the style called Greek revival. In the 19th century, English architect Sir Joseph Paxton created the Crystal Palace (1850-1851) in London, a vast exhibition hall which foreshadowed industrialized buildings and the widespread use of cast iron and steel. Also important in its innovative use of metal was the great Eiffel Tower (1887-1889) in Paris.

Modern Architecture

At the beginning of the 20th century, some designers refused to work in borrowed styles. Spanish architect Antoni Gaudí was the most original; his Casa Milá (1905-1907) and the unfinished Church of the Holy Family (1883-1926) in Barcelona exhibit a search for new organic structural forms. His work has some relation with the movement called art nouveau, which had begun in Brussels and Paris.

American architect Louis Sullivan gave new expressive form to urban commercial buildings. His career converges with the so-called Chicago school of architects, whose challenge was to invent the skyscraper or high-rise building. An apprentice of Sullivan's, Frank Lloyd Wright, became America's greatest native architect.

In Germany, the Bauhaus school brought together architects, painters, and designers from several countries, all determined to formulate goals for the visual arts in the modern age. Its first director was Walter Gropius; its second was Ludwig Mies van der Rohe. Gropius, and Mies eventually established themselves in the United States, where they were influential as architects and teachers. Their contemporary, Swiss-French architect Le Corbusier, also played great on modern architecture.

The style initiated by the Bauhaus architects and termed the International style gradually prevailed after the 1930s. The elegance and subtle proportions of Mies' Lake Shore Drive Apartments (1951) in Chicago and the Seagram Building (1958) in New York City represent modernism at its finest.

Between about 1965 and 1980, architects and critics began to support postmodernism. Although postmodernism is not a cohesive movement based on a distinct set of principles, in general postmodernists value individuality, intimacy, complexity, and occasionally humor.

By the early 1980s, postmodernism had become the dominant trend in American architecture and an important phenomenon in Europe as well. Its success in the United States owed much to the influence of Philip C. Johnson (as did modernism 50 years earlier). His AT&T Building (1984) in New York City immediately became a landmark of postmodern design.

VOCABULARY

New Words

Islam *n.* 伊斯兰教
mosque *n.* 清真寺
ruin *n.* 废墟，遗迹
ornamented *a.* 装饰的
floral *a.* 花的，花图案的
Arabic calligraphy 阿拉伯字体书写
principle *n.* 法则，原则，原理
threshold *n.* 起点，开端
oval *a.* 椭圆形的
regal *a.* 宏伟的，豪华的，适合帝王的
prominence *n.* 显著，杰出
exuberant *a.* 充满活力的
patron *n.* 主顾，业主，资助人
prototype *n.* 原型
foreshadow *v.* 预示，预兆，预先给予暗示
organic *a.* 有机的，有机物的

New Words

converge *v.* 会聚，汇合
formulate *v.* 规划，设计，构想
initiate *v.* 发起，开始
prevail *v.* 流行，盛行
critic *n.* 批评家，评论家
cohesive *a.* 有关联的
individuality *n.* 个性，独特性
intimacy *n.* 亲密性，隐私
complexity *n.* 复杂性

Architectural Terms

motif *n.* 基本图案，基本色彩
glazed tile 釉面砖
marble *n.* 大理石
arabesque *n.* 阿拉伯式花饰（由花果、蔓藤、几何图形等组成的精细图案，用作地毯花纹、壁画装饰等）
mannerism 风格主义（16世纪欧洲的一种艺术风格，强调形式的奇巧，风格上的个人癖好或对别人独特风格、技法的搬用和模仿；亦称"矫饰主义"或"体裁主义"）
entablature *n.* 古典柱式的顶部（由过梁、雕带和挑檐三部分组成）
recession *n.* 墙壁等的凹处，缩进处
neoclassicism 新古典主义（18世纪末、19世纪初流行的一种崇尚庄重典雅的建筑风格）
asymmetry *n.* 不对称
cast iron 铸铁
pediment *n.* 山花，三角墙（指希腊式古典建筑正面门廊顶上的装饰性三角形山头）
art nouveau 新艺术（约1890–1910间流行于欧洲和美国的一种装饰艺术风格，以曲折有致的线条为其特色）
modernism 现代主义建筑风格
postmodernism 后现代主义建筑风格

Proper Names

Sāmarrā 撒马拉（伊拉克中北部城市，位于巴格达西北偏北部的底格里斯河畔）
Córdoba 科尔多瓦（西班牙城市，曾经为阿拉伯倭马亚王朝都城）
Alhambra 爱尔罕布拉宫（古迹，在西班牙南部城市格拉纳达，是13–14世纪时的摩尔人宫殿）
Granada 格拉纳达（西班牙南部城市，位于科尔多瓦东南，由摩尔人于8世纪创建）
Tomb of Humayun 胡马雍之陵（位于印度，胡马雍为莫卧儿帝国皇帝）
Delhi 德里（印度中部偏北城市，自古以来该市就占重要地位，17世纪沙·贾汗皇帝重建这座古城围住莫卧儿皇宫）
Agra 阿格拉（印度北部城市，泰姬陵所在地）
Filippo Brunelleschi 菲利波·布鲁内莱斯基（1377–1446，意大利文艺复兴初期建筑师，代表作有圣洛伦佐教堂和弗洛伦萨的圣马利亚教堂）
Florence Cathedral 佛罗伦萨大教堂
Saint Peter's Basilica 圣彼得大教堂（教皇庭，天主教的中心，坐落在梵蒂冈城）
Vatican City 梵蒂冈（罗马教廷所在地，在

意大利罗马城西北角的梵蒂冈高地上)
Michelangelo 米开朗琪罗(1475–1564, 意大利文艺复兴盛期成就卓著的雕刻家、画家、建筑师, 建筑设计罗马圣彼得大教堂圆顶)
Baldassare Peruzzi 巴尔达萨雷·佩鲁齐(1481–1536, 意大利文艺复兴盛期建筑师和画家, 1520年被任命为罗马圣彼得大教堂的建筑师)
Giulio Romano 朱利奥·罗马诺(1499? –1546, 意大利文艺复兴晚期画家和建筑家, 拉斐尔的学生, 意大利风格主义奠基人, 曾设计圣贝内德托教堂)
Giacomo da Vignola 贾科莫·达·维尼奥拉(1507–1573, 意大利建筑大师, 风格主义建筑代表之一, 其罗马耶稣会教堂的十字形设计对后世的教堂建筑有很大的影响, 著有《建筑五柱式规范》等)
Gianlorenzo Bernini 乔洛伦佐·伯尼尼(1598–1680, 意大利雕塑家、画家和建筑家, 意大利巴洛克风格的杰出代表, 以其流畅、动感的雕塑及其为包括圣彼得大教堂在内的许多教堂的设计而著称)
Louis XIV 路易十四(1638–1715, 法国国王, 1643–1715年在位, 绰号太阳王)
Château de Versailles (法国)凡尔赛宫(欧洲最大王宫, 凡尔赛为法国中北部一个城市, 位于巴黎西南以西)
Sir Christopher Wren 克里斯托弗·雷恩爵士(1632–1723, 英国著名建筑师, 1666年伦敦大火后设计圣保罗教堂等50多座伦敦教堂, 还有许多宫廷建筑、图书馆等)
Rococo style 洛可可式(18世纪初起源于法国、18世纪后半期盛行于欧洲的一种建筑装饰艺术风格, 其特点是精巧、繁琐、华丽)
Bavaria 巴伐利亚(州)(德国南部地区, 原为公爵领地)
Austrian Benedictine Abbey 奥地利本笃会修道院(本笃会为基督教分支教派团体)
Ottobeuren 奥托伯伦堡(奥地利城市)
Federal style 美国联邦时期风格
Industrial Revolution 工业革命(发生于18世纪末期英国)
Sir Joseph Paxton 约瑟夫·帕克斯顿爵士(英国著名建筑师)
Greek revival (19世纪上半叶盛行的)希腊复兴式
Crystal Palace (伦敦)水晶宫(第一届世界博览会会址)
Eiffel Tower (巴黎)埃菲尔铁塔(在塞纳河南岸)
Antoni Gaudí 安东尼·高迪(1852–1926, 西班牙著名建筑师, 被尊为有机主义建筑流派开创人)
Casa Milá 米拉公寓
Church of the Holy Family 神圣家族大教堂
Barcelona 巴塞罗那(西班牙东北部港市)
Brussels 布鲁塞尔(比利时首都和最大城市)
Chicago school 芝加哥学派(美国建筑研究学派, 或指其风格)
Walter Gropius 沃尔特·格罗皮厄斯(1883–1969, 德裔美国著名建筑师, 首创金属构架玻璃悬强建筑)
Le Corbusier 勒·科比西埃(1887–1965, 瑞士裔法国建筑师, 重要作品有马赛公寓等)
Lake Shore Drive Apartments 滨湖大道公寓楼(雷克·肖大道公寓楼, 在芝加哥)
Seagram Building 西格拉姆大厦(纽约)
AT&T Building 美国电话电报公司大厦(纽约)

EXERCISES

I. Translate the following English into Chinese and Chinese into English.

1. marble ______________
2. mosque ______________
3. cast iron ______________
4. glazed tile ______________
5. motif ______________
6. 古典柱式的顶部 ______________
7. 流行，盛行 ______________
8. 洛可可式 ______________
9. 新古典主义 ______________
10. 山墙，山花 ______________

II. Decide whether the following statements are True or False.

1. Islamic architecture once upon a time spread to India and Spain.
2. The cultural revolution in Western Europe now called the Renaissance brought a counteraction against the principles and styles of ancient Greek and Roman architecture.
3. Toward the mid-16th century such leading Italian architects as Michelangelo, Baldassare Peruzzi, Giulio Romano, and Giacomo da Vignola began to use the classical Roman elements in ways that became known as mannerism.
4. Italians, French and English were the pioneers of Baroque style.
5. Spanish architect Antoni Gaudí's work has nothing to do with the movement called art nouveau, which had begun in Brussels and Paris.
6. Though postmodernism was not a cohesive movement, by the early 1980s, it had become the dominant trend in American architecture and an important phenomenon in Europe as well.

III. Choose the best answer to each of the following questions.

1. One of the most important forms of Islamic architecture is ______.

   A. church　　B. mosque　　C. palace　　D. castle

2. Italian architect _________who designed a dome for the Florence Cathedral stood at the threshold between Gothic and Renaissance.

   A. Michelangelo　　B. Baldassare Peruzzi

   C. Giulio Romano　　D. Filippo Brunelleschi

3. Famous architect Sir Christopher Wren whose masterpiece is Saint Paul's Cathedral is a(n) _______.

   A. Italian　　B. French　　C. English　　D. German

4. The _______ foreshadowed industrialized buildings and the widespread use of cast iron and steel.
   A. The Crystal Palace B. Eiffel Tower
   C. Church of the Holy Family D. AT&T Building
5. _______ was the first director of Bauhaus school.
   A. Frank Lloyd Wright B. Ludwig Mies van der Rohe
   C. Le Corbusier D. Walter Gropius

IV. Oral task

In next class, you'll be asked to give an oral report based on one of the following questions. Work in teams and search the library or Internet for relevant pictures, facts or stories to support your points.

→ 1 What is the major form of Islamic architecture and what are the features of it?
→ 2 What changes did the Renaissance bring to architecture?
→ 3 Can you name some modern architectural styles and their associated architects?

TEXT C

The Brief History of Landscape Architecture

Landscape architecture is the science and art of modifying land areas by organizing natural, cultivated, or constructed elements according to a comprehensive, aesthetic plan. These elements include topographical features such as hills, valleys, rivers, and ponds; growing things such as trees, shrubbery, grass, and flowers; and constructions such as buildings, terraces, and fountains. Landscape architecture covers a wide area of concerns, ranging from the setting out of small gardens to the ordering of parks, malls, and highways. A landscape architect, who usually holds an academic degree in this field, designs most of the plans.

Principles

Working alone or with a town planner, traffic engineer, or building architect, as the project requires, the landscape architect considers the proposed use for the site, the layout of the terrain, climate and soil conditions, and costs. An overall plan is established, taking into account proportion and scale, as well as natural land formations. The landscape architect also considers contrasts in shady masses and open, sunny spaces, especially in relation to the climate. Contrasts in the size, color, and texture of plant material are also important. Planting may be designed according to seasons so that different parts of a garden bloom at different times.

Ancient World

In Mesopotamia, the Hanging Gardens of Babylon were one of the Seven Wonders of the World. They included full-size trees planted on earth-covered terraces in a corner of the palace complex of King Nebuchadnezzar II. The Assyrians and Persians developed tree-filled parks for hunting on horseback. In ancient Greece, sacred groves were preserved as the habitats of divinities, and houses included a walled court or garden.

Non-Western World

Islamic gardens were usually one or more enclosed courts surrounded by cool arcades, planted with trees and shrubs, and enhanced with colored tilework, fountains, and pools. The Moors in Spain built such gardens at Córdoba, Toledo, and especially at the Alhambra in Granada.

In China, palaces, temples, and houses were built around a series of courtyards, which might include trees and plants. Japan has a long tradition of garden settings closely integrated with the buildings. The gardens traditionally have included pools and waterfalls; rocks, stone, and sand; and evergreens. Every element of a garden is carefully planned to create an effect of restraint, harmony, and peace.

Medieval, Renaissance, and Baroque Periods

In medieval Europe, which was ravaged by invasions and incessant wars, gardens were generally small and enclosed for protection within the walls of monasteries and castles. During the Renaissance in Italy, castles gave way to palaces and villas with extensive landscaped grounds. A 15th-century example is the garden of the Medici Villa, in Florence.

Italian gardens of the 17th century became more complex in the dramatic Baroque style. They were distinguished by lavish use of serpentine lines, groups of sculptured allegorical figures, and multiple fountains and waterfalls. However, in the 17th century, France replaced Italy as the primary inspiration of architectural and landscape design. The great châteaus of the Loire Valley, such as Chambord, were laid out with formal gardens and with extensive forested parks. The vast building programs of King Louis XIV included miles of symmetrically arranged gardens, such as the royal gardens at Versailles.

Romantic Period

In the late 18th century the rise of romanticism led to important changes in landscape architecture as well as in other arts. Architects began to imitate nature rather than restrain it. English landscape architect Humphry Repton created the so-called English garden, believing that a house was best set off by formal flower beds that subtly merged into the background.

The first major public example of landscape architecture in the United States was Central Park in New York City, designed in 1857 by Frederick Law Olmsted and Calvert Vaux. It was so successful that it influenced the nationwide creation of public parks. The profession of landscape architecture, as distinct from architecture and horticulture, was established largely through the success of Olmsted and Vaux.

20th Century

In the early 20th century, landscape architects concentrated on integrating houses with their surroundings, but the worldwide economic depression of the 1920s and 1930s forced a shift to large-scale public works, with landscape architects and planners collaborating on communities, regional areas, and vast state and national projects. The proliferation of shopping malls, new suburbs, revitalized urban cores, and new educational facilities has given 20th-century landscape architects unparalleled opportunities to refine their art and to create new forms.

VOCABULARY

New Words

cultivated *a.* 人工栽培的
topographical *a.* 地形学的
shrubbery *n.* 灌木丛
terrace *n.* 梯田
academic *a.* 学术上的
formation *n.* 形状，结构
sacred *a.* 神圣的
grove *n.* 小树林，园林
habitat *n.* 栖息地，居住地
divinity *n.* 神，神氏
ravage *v.* 破坏，蹂躏
invasion *n.* 入侵
incessant *a.* 不断的
monastery *n.* 修道院
lavish *a.* 奢侈的，过度的
serpentine *a.* 蜿蜒的
allegorical *a.* 寓言的，讽喻的
set off 衬托出
collaborate *v.* 协作，合作
proliferation *n.* 迅速增长，扩散
shopping mall 商业街，商场，购物中心
revitalized *a.* 新生的
unparalleled *a.* 无比的，空前的

Architectural Terms

landscape architecture 景观建筑，景观营造
complex *n.* 建筑群
arcade *n.* 拱廊
tilework *n.* 砖瓦工艺
château *n.* 法国式城堡、宫殿
horticulture *n.* 园艺（学）

Proper Names

Persian 波斯人（古代或中世纪）

Moor 摩尔人（非洲西北部阿拉伯人与柏柏尔人的混血后代，于公元8世纪进入并统治西班牙）

Toledo 托莱多（西班牙中部城市，位于马德里西南偏南，曾经是摩尔人的首都）

Medici Villa 美第奇家宅（美第奇是意大利文艺复兴时期统治佛罗伦萨的贵族，曾赞助过米开朗琪罗等艺术家）

Loire Valley 卢瓦尔河谷（卢瓦尔河是法国最长的河流，位于法国中部）

Chambord 商堡（法国中北部一村镇，以文艺复兴时期法王弗朗西斯一世在此兴建的壮观华美的城堡而闻名）

Humphry Repton 亨弗利·雷普顿（18世纪英国著名景观设计师）

Frederick Law Olmsted 弗雷德里克·劳·奥姆斯特德（1822–1903，美国景观设计师，纽约市中央公园的主设计师）

Calvert Vaux 卡尔弗特·沃克斯（1824–1895，英裔美国景观设计师，纽约城内的中心公园的设计者之一）

Answer the following questions according to the text.

1. How do you understand landscape architecture?
2. What are the principles of landscape architecture practice?
3. What are the features of landscape architecture in the non-Western world?
4. What were the features of landscape architecture during the Medieval, Renaissance and Baroque periods?
5. What changes did landscape architecture undergo during the Romantic period?

3
Unit

Architectural Styles

Warming Up

1. If you were an architect, would architectural style be the first thing you think about when designing a building?
2. Work in groups to talk about the main distinctions between the Western architectural styles and the Eastern ones.

TEXT A

Baroque Style

Emerging in both Rome and Paris shortly after 1600, the Baroque in art and architecture soon spread throughout Europe, where it prevailed for one hundred and fifty years. During this period new social and political systems resulted in the concentration of power in the hands of individuals with absolute authority. Architecture affirmed this—through the structures and decorative programs of palaces, churches, public and government buildings, scientific and commercial buildings, and military installations. Magnificent churches, fountains, and palaces attested to the renewed strength of the Popes in Rome, while architects also gave new forms to churches for the Protestant and Russian Orthodox liturgies. Baroque architects had been schooled in the classical Renaissance tradition, emphasizing symmetry and harmonious proportions, but their designs revealed a new sense of dynamism and grandeur. Renaissance architects had sought to engage the intellect, with their focus on divine sources of geometry, while their successors aimed to overwhelm the senses and emotions. Baroque architects also mastered the unification of the visual arts—painting, sculpture, architecture, garden design, and urban planning—to a remarkable degree, producing buildings and structures with a heightened sense of drama and power.

Baroque and Bernini

Buildings of the period are composed of great curving forms with undulating facades, ground plans of unprecedented size and complexity, and domes of various shapes, as in the churches of Francesco Borromini, Guarino Guarini, and Balthasar Neumann. Many works of Baroque architecture were executed on a colossal scale, incorporating aspects of urban planning and landscape architecture. This is most clearly seen in Bernini's elliptical piazza in front of St. Peter's in Rome.

Bernini was an architect, a painter, and a sculptor—one of the most important and imaginative artists of the Italian Baroque era and its most characteristic and sustaining spirit. The design of the monumental piazza in front of St. Peter's was Bernini's largest and most impressive single project. Bernini had to adjust his design to some preexisting structures on the site—an ancient obelisk the Romans had brought from Egypt and

▲ view from the roof of St. Peter's, Rome

▲ Trevi Fountain, Rome

a fountain Maderno designed. He used these features to define the long axis of a vast oval embraced by colonnades joined to the St. Peter's facade by two diverging wings. Four rows of huge Tuscan columns make up two colonnades, which terminate in severely classical temple fonts. The dramatic gesture of embrace the colonnades make as viewers enter the piazza symbolizes the welcome the Roman Catholic Church gave its members during the Counter-Reformation. Bernini himself referred to his design of the colonnade as appearing like the welcoming arms of the church. By its sheer scale and theatricality, the complete St. Peter's fulfilled Catholicism's needs in the 17 century by presenting an awe-inspiring and authoritative vision of the Church.

(*The use of decorative gardens and fanciful and elaborate fountains was also typical of the Italian Baroque. Various estates and city plazas in Rome, for example, testify to this Baroque fashion for highly ornate and creatively sculpted fountains.*)

Division of the Baroque Period

For convenience the Baroque period is divided into three parts:

Early Baroque, c.1590–c.1625

The early style was preeminent under papal patronage in Rome where Carracci and Caravaggio and his followers diverged decisively from the artifice of the preceding mannerist painters. Bernini abandoned an early mannerism in his sculpture, allowing him to express a new naturalistic vigor. In architecture, Carlo Maderno's facades for St. Susanna and St. Peter's moved toward a more sculptural treatment of the classical orders.

High Baroque, c.1625–c.1660

The exuberant trend in Italian art was best represented by Bernini and Borromini in architecture, by Bernini in sculpture, and by Da Cortona in painting. The classicizing mode characterized the work of the expatriate painters Poussin and Claude Lorrain. This period produced an astonishing number and variety of international painters of the first rank, including Rembrandt, Rubens, Velázquez, and Anthony van Dyck.

Late Baroque, c.1660–c.1725

During this time Italy lost its position of artistic dominance to France, largely due to the patronage of Louis XIV. The late Baroque style was especially popular in Germany and Austria, where many frescoes by the Tiepolo family were executed. The extraordinarily theatrical quality of the architecture in these countries is best seen in the work of Neumann and Johann Bernhard Fischer von Erlach. From Europe the Baroque spread across the Atlantic Ocean to the New World. Gradually the massive forms of the Baroque yielded to the lighter, more graceful outlines of the rococo.

VOCABULARY

New Words

affirm *v.* 证实
installation *n.* 装置，设备
attest to 证明，表明
Pope *n.* 教皇
liturgy *n.* 礼拜仪式
symmetry *n.* 对称
divine *a.* 神的
unification *n.* 合一，联合
undulating *a.* 呈波浪形的
unprecedented *a.* 无前例的，前所未闻的
elliptical *a.* 椭圆形的
sustaining *a.* 用以支撑的
diverging *a.* 分散的，分开的
terminate *v.* 结尾
sheer *a.* 纯粹的
theatricality *n.* 夸张，做作
preeminent *a.* 卓越的
papal *a.* 教皇的
patronage *n.* 资助，赞助
expatriate *a.* 生活在国外的，被流放的

Architectural Terms

facade *n.* （建筑物的）正面
colossal *a.* （柱型）高大的
piazza *n.* （尤指意大利等城市中的）露天广场
obelisk *n.* 方尖碑
colonnade *n.* 列柱，柱廊
Tuscan *a.* 托斯卡纳柱型的（古罗马建筑中的一种柱子的式样）
fresco *n.* 湿壁画

Proper Names

Protestant 新教徒
Russian Orthodox 俄罗斯东正教
Francesco Borromini 弗朗切斯科·博罗米尼(1599–1677, 意大利建筑家)
Guarino Guarini 加里诺·加里尼(1624–1683, 意大利建筑家)
Balthasar Neumann 约翰·巴塔萨·纽曼(1687–1753, 德国建筑师)
Carlo Maderno 卡罗·玛丹诺(意大利建筑师, 以巴洛克风格著称)
Carracci 卡拉齐画派(他们的作品影响并导致巴洛克式艺术过渡期的风格主义的变革)
Caravaggio 米开朗琪罗·卡拉瓦乔
St. Susanna 圣苏珊纳大教堂
Da Cortona 达·科托纳(意大利建筑师、画家, 巴洛克艺术的倡导者)
Poussin 尼古拉斯·普珊(1594–1665, 法国画家)
Claude Lorrain 克劳德·洛兰(1600–1682, 法国风景画家, 开创表现大自然诗情画意新风格)
Rembrandt 伦勃朗(1609–1669, 荷兰著名画家, 以肖像画著名)
Rubens 彼得·保罗·鲁本斯(1577–1640, 比利时画家, 巴洛克艺术代表)
Velázquez 韦拉兹奎兹(1599–1660, 西班牙画家, 画风写实)
Anthony van Dyck 安东尼·凡·戴克(1599–1641, 比利时画家)
Tiepolo 乔瓦尼·巴蒂斯特·泰波罗(1696–1770, 意大利画家)
Johann Bernhard Fischer von Erlach 约翰·伯恩哈德·菲歇尔·冯·厄拉策(1656–1723, 奥地利建筑师)

EXERCISES

I. Match the Chinese expressions with their English equivalents.

1. 方尖碑
2. 喷泉
3. 列柱
4. 建筑物的正面
5. 露天广场
6. 对称
7. 托斯卡纳柱型的
8. 视觉艺术
9. 园林景观建筑
10. 城市规划

A. symmetry
B. visual arts
C. obelisk
D. piazza
E. urban planning
F. landscape architecture
G. fountain
H. Tuscan
I. colonnade
J. facade

II. Decide whether the following statements are True or False.

1. Although Baroque architects' designs revealed a new sense of dynamism and grandeur, they had been rooted in classical Renaissance tradition, emphasizing symmetry and harmonious proportions.
2. Baroque architects' talent and capacity had nothing to do with visual arts such as painting, sculpture, garden design, and urban planning.
3. In designing the magnificent oval plaza in front of St. Peter's, Bernini had to adjust his design to some preexisting structures on the site.
4. In early Baroque period, Bernini stuck to an early mannerism in his sculpture, allowing him to express a new naturalistic vigor.
5. High Baroque period produced not only great architects like Bernini and Borromini but also an astonishing number and variety of international painters including Da Cortona, Rembrandt, Rubens, Velázquez, and Anthony van Dyck.

III. Choose the best answer to each of the following questions.

1. Buildings of Baroque period are composed of great curving forms with ________ facades, ground plans of unprecedented size and complexity, and domes of various shapes.
   A. plain　　B. waving　　C. broad　　D. subtle
2. Four rows of huge _______ columns make up two colonnades, which form the enclosure and terminate in classical temple fonts.
   A. Doric　　B. Ionic　　C. Corinthian　　D. Tuscan
3. The dramatic gesture of embrace that the colonnades make as viewers enter the St. Peter's Piazza symbolizes the welcome the ________ Church gave its members during the Counter-Reformation.
   A. Eastern Orthodoxy　　B. Protestant　　C. Roman Catholic　　D. Puritan
4. During late Baroque, Italy lost its position of artistic dominance to ________, largely due to the patronage of Louis XIV and subsequently the late Baroque style was especially popular in _______.
   A. France; Germany and Austria　　B. Britain; Germany and Austria
   C. France; Russia and America　　D. Britain; Russia and America
5. Gradually the massive forms of Baroque style evolved into a lighter, more graceful outlines of the _______.
   A. classicism　　B. neoclassicism　　C. rococo　　D. eclecticism

IV. Oral task

In next class, you'll be asked to give an oral report based on one of the following questions. Work in teams and search the library or Internet for relevant pictures, facts or stories to support your points.

→ 1 What is your understanding of the Baroque style in architecture?

→ 2 Introduce a world-famous architecture of Baroque style.

TEXT B

Architecture and Decorative Art of Neoclassicism

▲ Château de Versailles

The neoclassical style developed following the excavation of the ruins of the Italian cities of Herculaneum in 1738 and Pompeii in 1748, the publication of such books as *Antiquities of Athens* (1762) by the English archaeologists James Stuart and Nicholas Revett, and the 1806 arrival in London of the Elgin Marbles. Extolling the "noble simplicity and calm grandeur" of Greco-Roman art, the German art historian Johann Winckelmann urged artists to study and "imitate" its timeless, ideal forms. His ideas found enthusiastic reception within the international circle of artists gathered about him in the 1760s in Rome.

Before the discoveries at Herculaneum, Pompeii, and Athens, only Roman classical architecture had been generally known, largely through the architectural etchings of classical Roman buildings of the Italian artist Giovanni Battista Piranesi. The new archaeological finds extended architecture's formal vocabulary, and architects began advocating buildings based on Greco-Roman models.

The work of the Scottish architect and designer Robert Adam, who in the 1750s and 1760s redesigned a number of stately English houses (among others, Syon House, 1762-1769, and Osterley Park, 1761-1780), introduced the neoclassical style to Great Britain. The Adam style, as it became known, remained somewhat rococo in its emphasis on surface ornamentation and refinement of scale, even as it adopted the motifs of antiquity.

In France, Claude Nicholas Ledoux designed a pavilion (1771) for the Comtesse du Barry at Louveciennes and a series of city gates (1785-1789) for Paris—structures that are exemplars of the earlier phase of neoclassical architecture; his later works, however, consisted of projects (never executed) for an ideal city in which the designs for buildings are frequently reduced to unadorned geometric shapes. After Napoleon became emperor in 1804, his official architects Charles Percier and Pierre François Léonard Fontaine worked to realize his wish to remake Paris into the foremost capital of Europe by adopting the intimidating opulence of Roman imperial architecture. The Empire style in architecture is epitomized by such mammoth public works as the triumphal arches at the Carrousel du Louvre, designed by Percier and Fontaine and begun in 1806, and the Champs-Élysées, designed by Jean-François Chalgrin and begun the same year. These works were far different in spirit from the visionary work of Ledoux.

Greek-inspired architecture in England is exemplified by such constructions as the Bank of England rotunda (1796) by Sir John Soane and the British Museum portico (1823-1847) by Sir Robert Smirke. The Greek revival was modified by the Regency style, notable architectural examples of which are the facades designed by John Nash for Regent Street (begun 1812) in London and his Royal Pavilion at Brighton (1815-1823). The neoclassical architecture of Edinburgh, Scotland, remained pristine, however, and earned that city the name the Athens of the North. Elsewhere, neoclassical architecture is exemplified in the work of the German Karl Friedrich Schinkel, such as the Royal Theater (1819-1821) in Berlin.

In the United States, one aspect of neoclassicism, the Federal style, flourished between 1780 and 1820. Based on the work of Robert Adam, it is exemplified in the work of Charles Bulfinch (Massachusetts State House, Boston, completed 1798). Thomas Jefferson studied the Maison-Carrée, a 1st-century Roman temple in Nîmes, France, and used it as a model for the State Capitol Building in Richmond, Virginia (1785-1789). Through his readings and travels, Jefferson developed a profound understanding of Roman architecture and applied his knowledge to the designs for his own home, Monticello; the University of Virginia campus; and preliminary contributions to the plans for the new national capital of Washington, D.C. Jefferson's work exemplifies neoclassical style in the United States.

The Greek revival style, based on 5th-century BC Greek temples and inspired by the Elgin Marbles, flourished during the first half of the 19th century in the United States.

The Second Bank of the United States (Philadelphia, 1824), designed by William Strickland, was influenced by a Doric temple. Both the Federal and Greek revival styles helped a young United States define its own architectural ethos.

▲ The Schonbrunn Palace, Vienna

The neoclassical style pervaded every type of decorative art. By the early 1760s Robert Adam's furniture designs revealed Greco-Roman motifs. Introduced into France, his simple, classical style became known as style étrusque (Etruscan style), favored by the court of Louis XV. With further adaptations of classical design, based on later archaeological finds, it evolved into the elegant style known as Louis XVI, favored by the royal family during the 1780s. Greek vases found in excavations became models for new types of ceramics: Wedgwood jasperware (for which Flaxman did many designs) in England and Sèvres porcelain in France.

Under Napoleon, former royal residences were redecorated for official use according to plans devised by Percier and Fontaine that included furniture, porcelain, and tapestries, all incorporating Greco-Roman design and motifs. Taken as a whole, such design complexes defined the Empire style in the decorative arts, and the style was soon emulated throughout Europe.

VOCABULARY

New Words

antiquity *n.* 古迹，古物
archaeologist *n.* 考古学家
extoll *v.* 赞美，颂扬
stately *a.* 庄严的，宏伟的
opulence *n.* 豪华，富裕
mammoth *a.* 巨大的
pristine *a.* 原始状态的，本来的
ethos *n.* 社会（或民族等）的精神特质
jasperware *n.* 碧玉细炻器
porcelain *n.* 瓷器，瓷
tapestry *n.* 织锦，挂毯
emulate *v.* 仿效

Architectural Terms

etching *n.* 蚀刻版画，蚀刻术
unadorned *a.* 未装饰的，朴实的
rotunda *n.* 圆形建筑，圆形大厅
portico *n.* (有圆柱的) 门廊，柱廊
ceramics *n.* 制陶术，制陶业

Proper Names

Herculaneum 海格力古城 (意大利南部古城)
Pompeii 庞培古城 (意大利南部古城，公元79年火山爆发，全城淹没)
James Stuart 詹姆士·斯图亚特
Nicholas Revett 尼古拉斯·利维特
Elgin Marbles 埃尔金大理石雕 (指一些雅典雕刻及建筑残件，于19世纪由英国伯爵Thomas Elgin运至英国，现藏不列颠博物馆)
Johann Winckelmann 约翰·温科尔曼
Giovanni Battista Piranesi 乔万尼·巴迪斯塔·皮兰尼西
Robert Adam 罗伯特·亚当
Syon House 西昂之宅
Osterley Park 奥斯特雷公园
Claude Nicholas Ledoux 克劳德·尼古拉斯·雷多
Comtesse du Barry 巴里伯爵夫人
Louveciennes 卢维谢纳 (法国城市)
Carrousel du Louvre 罗浮宫旋转木马
Champs-Élysées 爱丽舍宫 (巴黎古建筑，法国总统官邸，建于18世纪初)
Jean-François Chalgrin 让·弗朗索瓦·查尔金
Sir John Soane 约翰·索恩爵士
the British Museum 大英博物馆
Sir Robert Smirke 罗伯特·斯莫克爵士
Regency style (英国) 摄政时期风格
John Nash 约翰·纳什
Regent Street 摄政大道 (伦敦)
Royal Pavilion at Brighton 布赖顿皇家观景阁
Brighton 布赖顿 (英国英格兰东南部城市)
Edinburgh 爱丁堡 (英国苏格兰首府)
Karl Friedrich Schinkel 卡尔·弗莱德里西·辛克尔
the Royal Theater in Berlin 柏林皇家剧院
Charles Bulfinch 查尔斯·布尔芬奇
Massachusetts State House 马萨诸塞州议会大厦
Thomas Jefferson 托马斯·杰斐逊 (1743–1826，美国第三任总统)
the Maison-Carrée 伽黑之宅
Nîmes 尼斯 (法国南部城市)
the State Capitol Building 美国国会大厦
Richmond 里士满 (美国弗吉尼亚州首府)
Virginia 弗吉尼亚州
Monticello 蒙蒂塞洛 (美国弗吉尼亚州中部夏洛茨维尔东南一住宅区，由托马斯·杰斐逊设计)
Philadelphia 费城 (美国宾西法尼亚州东南部港市)
William Strickland 威廉·斯催克兰
Etruscan (意大利中西部古国) 伊特鲁里亚的，伊特鲁里亚人的
Wedgwood 韦奇伍德装饰陶瓷 (商标名称)
Flaxman 弗拉克斯曼 (1755–1826，英国雕刻家)
Sèvres 塞夫勒 (法国北部城市)

EXERCISES

I. Translate the following English into Chinese and Chinese into English.

1. neoclassical style ______________
2. portico ______________
3. excavation ______________
4. etching ______________
5. geometric shape ______________
6. 织锦, 挂毯 ______________
7. 制陶术， 制陶业 ______________
8. 建筑气韵 ______________
9. 圆形建筑， 圆形大厅 ______________
10. 古物, 古迹 ______________

II. Match the following designers with their works.

designer

1. Jean-François Chalgrin
2. Percier and Fontaine
3. Robert Adam
4. Karl Friedrich Schinkel
5. Claude Nicholas Ledoux
6. Charles Bulfinch
7. Sir John Soane
8. William Strickland
9. Sir Robert Smirke

works

A. Syon House
B. a pavilion for the Comtesse du Barry at Louveciennes
C. the triumphal arches at Carrousel du Louvre
D. the Champs-Élysées
E. the Bank of England rotunda
F. the British Museum portico
G. the Royal Theater
H. Massachusetts State House
I. the Second Bank of the United States

IV. Oral task

In next class, you'll be asked to give an oral report based on one of the following questions. Work in teams and search the library or Internet for relevant pictures, facts or stories to support your points.

➔ 1 What are the differences between neoclassic style and Baroque style?

➔ 2 Besides architecture and decorative art, can you exemplify other forms of art of neoclassicism?

TEXT C

Chinese Classical Architectural Style

China is a country with a long history. On this land, our ancestors left us an abundance of splendid, time-honored architectural legacy, which has undergone thousands of years of development to become a distinct part of world architectural history. These features are demonstrated mainly in the following aspects:

1. Wooden Frameworks

Wooden frameworks for buildings appeared at a very early period of Chinese history. First, rows of wooden pillars are raised from the ground, on which horizontal wooden roof beams and crossbeams are placed. The roof timbers are laid on the beams, so that the weight of the roof is all transmitted to the ground by way of the beams and the upright pillars. The advantages of this form of structure are as follows: first, the wooden framework bears all the weight of the building, which makes the installation of both the outer and inner walls flexible and able to be placed in accordance with practical needs. Doors and windows can be installed between the erected pillars, or the pillars can be left open. Inside, the house can be divided into spaces with different purposes using wooden partitions and screens. Second, the wooden framework is shock-resistant, because the parts are linked by mortise-and-tenon joint. So, when subjected to a violent shock such as that from an earthquake, a wooden framework is less likely to break or fall down than a brick or stone one. Third, a wooden framework is easy to construct. Wood is a natural material, not like bricks and tiles which are manufactured. Compared with stones which are also natural materials, wood is much easier to obtain, refine and work on.

2. Collective Layout of Structures

Traditional Chinese buildings are always found in pairs or groups, whether they are residences, temples or palaces. The Siheyuan (courtyard house or quadrangle) in Beijing is the typical form of residence in north China. It is a compound with houses around a square

courtyard. The main house in the courtyard is occupied by the head of the family, and the junior members live in the wings on each side. This layout not only conforms to the feudal Chinese family moral principle of distinction between the older and younger, and male and female members, but also provides a quiet and private environment for family life. Temples and palaces also sometimes display this layout. In the Forbidden City in Beijing, there are nearly 1,000 halls of varied sizes which are all grouped around large or small courtyards.

In garden architecture, in order to create an environment with hills and waters of natural beauty in a limited space, structures are usually carefully separated and laid irregularly to make variable spaces and different landscapes. Although occasionally grouped around courtyards, the pavilions, terraces, towers and halls are often separate scenes with a tenuous connection between them.

In both regular and irregular architectural complexes, decorative archways, pillars, screen walls, and stone lions and tablets besides small buildings play an important role in dividing space and forming scenes.

Most structures in traditional Chinese architecture are simple rectangles, and it is the architectural complex composed by single structures rather than the single structures themselves that expresses the broadness and magnanimousness of ancient Chinese architecture.

3 The Artistic Treatment of the Architectural Image

Ancient Chinese artisans ingeniously made the heavy roofs of buildings look light and graceful by forming the ridges and eaves into curves, and making the four corners stick up. The style of a roof can be divided into four basic types: fudian(wings), xieshan(hip and gable), xuanshan(suspended gable) and yingshan(hard gable), which denote a roof with a single layer, several layers, four corners and many corners. All these make the huge roof an important component of ancient Chinese architecture with an outstanding artistic image.

▲ upturned eaves

◀ *toukong*

The decorations on ancient Chinese structures have cultural connotations as well as esthetic ones. The dragon heads on the edges of roof ridges signify the spurting of water to douse fires. The dragon, phoenix, tiger and tortoise were regarded as sacred animals by the ancient Chinese, and they carved images of them on eaves tiles which were exclusively used on imperial structures. The emperors were supposed to be descendants of dragons, so there are images of dragons all over imperial structures, from balustrade column heads, terrace steps and stone foundations of pillars to roof beams, paintings on ceilings and carvings on doors and windows. Symbols denoting happiness, honor and longevity can be seen everywhere on traditional Chinese structures, including palaces, temples, gardens, residences, gateways, windows and roof beams. Bats represent happiness, deer stand for honor, and pines, cranes and peaches represent longevity. In addition, there are various patterns made by putting Chinese characters together, like the combination of the characters meaning happiness, longevity and ten thousand.

Ancient artisans were also good at using colors to decorate buildings. In the Forbidden City, stretches of yellow glazed tiles glitter under the blue sky, and there is a pleasing contrast between the dark green used beneath eaves, red doors, windows and walls, and white terrace foundations. Structures in private gardens in the south tend to have white walls, gray bricks and black structures create the simple and refined atmosphere beloved by the literati of old.

Wooden framework, collective layout and the artistic treatment of architectural image are the basic features of ancient Chinese architecture.

VOCABULARY

New Words

abundance *n.* 充裕, 丰富
time-honored *a.* 古老而受到尊重的
horizontal *a.* 水平的
partition *n.* 隔开物
quadrangle *n.* 四边形
magnanimousness *n.* 宽宏大量, 慷慨
ingeniously *adv.* 巧妙地
denote *v.* 指示
spurt *v.* 喷射
douse *v.* 在……上浇水
longevity *n.* 长寿
crane *n.* 鹤
glitter *v.* 闪闪发光
literati *n.* 文人学士

Architectural Terms

pillar *n.* 柱子
beam *n.* (建筑物等的) 横梁
timber *n.* 木材, 木料
mortise-and-tenon 榫眼与榫舌
tile *n.* 瓷砖
eaves *n.* 屋檐, 房檐
upturned eaves 尖端向上翻的檐, 挑檐
toukong 斗拱 (中国建筑特有的一种结构。在立柱和横梁交接处, 从柱顶上加的一层层探出成弓形的承重结构叫拱, 拱与拱之间垫的方形木块叫斗, 合称斗拱)
roof ridge 屋脊
balustrade *n.* 栏杆, 扶手

Answer the following questions according to the text.

1. What aspects did the features of Chinese classical architecture mainly demonstrate?
2. What are the advantages of the structure of wooden frameworks?
3. What is the typical form of residence in Beijing which presents the collective layout of structures?
4. What are the four basic types of roof?
5. What do the dragon heads on the edges of roof ridges signify?

Unit 4 Distinguished Architects

Warming Up

1. Do you know any world-famous architects and their works?
2. Would you choose to be an architect? Give your reasons.

TEXT A

Architecture is that great living creative spirit which from generation to generation, from age to age, proceeds, persists, creates, according to the nature of man, and his circumstances as they change. That is architecture.—Frank Lloyd Wright, 1937

The Life of Frank Lloyd Wright

Frank Lloyd Wright was born as Frank Lincoln Wright in Richland Center in southwestern Wisconsin, on June 8, 1867. His father, William Carey Wright, was a musician and a preacher. His mother, Anna Lloyd-Jones was a teacher. It is said that Anna Lloyd-Jones placed pictures of great buildings in young Frank's nursery as part of training him up from the earliest possible moment as an architect.

Wright briefly studied civil engineering at the University of Wisconsin in Madison, after which he moved to Chicago to work for a year in the architectural firm of J. Lyman Silsbee. In 1887, he hired on as a draftsman in the firm of Adler and Sullivan, run by Louis Sullivan (design) and Dankmar Adler (engineering). The Adler and Sullivan firm was just the right place to be for a young man aspiring to be a great architect, as it was at the leading edge of American architecture at the time. Wright eventually became the chief draftsman, and also the man in charge of the firm's residential designs. Under Sullivan, whom Wright called "Lieber Meister" (beloved master), Wright began to develop his own architectural ideas of interior space. Rejecting the existing view of rooms as single-function boxes, Wright created overlapping and interpenetrating rooms with shared spaces. He designated use areas with screening devices and subtle changes in ceiling heights and created the idea of defined space as opposed to enclosed space.

Wright started his own firm in 1893. Between 1893 and 1901, 49 buildings designed by Wright were built. During this period he began to develop his ideas which would come to together in his "Prairie House" concept—a long, low building with hovering planes and horizontal emphasis. He developed these houses around the basic crucifix, L or T shape and utilized a basic unit system of organization. He integrated simple materials such as brick, wood, and plaster into the designs. Into 1909, he developed and refined the prairie style. Frank Lloyd Wright founded the "Prairie school" of architecture, and his art of this early productive period in his life is also considered as part of the "Arts and Crafts Movement".

This very productive first phase in Wright's career ended in 1909, when he left his first wife and 5 children to go to Germany. He was joined there by his lover—Mamah Borthwick Cheney, the wife of a former client. From 1912-1914, Wright and Ms. Chaney lived together at Taliesin, his home and studio in Wisconsin. This period ended when a crazed servant murdered Ms. Chaney and 6 others, also setting a fire that destroyed much of Taliesin.

During the period from 1914-1932, a time of personal turmoil and change, architectural designs included the Imperial Hotel in Tokyo (a large and complex design that required much time in Japan to oversee it), and the concrete California residences. Few commissions were completed toward the end of this period, but Wright did lecture and publish frequently, with books including *An Autobiography* in 1932.

In 1932 Wright established the Taliesin Fellowship—a group of apprentices who did construction work, domestic chores, and design studies. *An Autobiography* served as an advertisement, inspiring many who read it to seek him out. The architect's output became more organized and prolific, with help of the numerous apprentices who assisted in design detail and site supervision. Four years later, he designed and built his most famous work, both Fallingwater and the Johnson Administration Building. These designs re-invigorated Wright's career and led to a steady flow of commissions, particularly for lower middle-income housing. Wright responded to the need for low income housing with the Usonian house, a development from his earlier prairie house.

Few buildings were produced during the war years, but the post-war period to the end of Wright's life was the most productive. He received 270 house commissions, and designed and built the Price Tower skyscraper, the Guggenheim Museum (which required Wright to live in New York City for a time), and the Marin County Civic Center.

▲ the Guggenheim Museum

Wright never retired; during the last part of his life, Wright produced a wide range of work. Particularly important was Taliesin West, a winter retreat and studio he built in Phoenix, Arizona. He died on April 9, 1959 at the age of ninety-two in Arizona. He was interred at the graveyard at Unity Chapel (which is considered to be his first building) at Taliesin in Wisconsin. The epitaph at his Wisconsin gravesite reads: "Love of an idea, is the love of God."

VOCABULARY

New Words

hire on 接受雇用
aspiring *a.* 有志气的，有抱负的
overlapping *a.* 重迭的，交搭的
interpenetrating *a.* 相互渗透的
designate *v.* 清楚地标出或指出
hovering *a.* 翱翔的，盘旋的
crazed *a.* 疯狂的
turmoil *n.* 混乱
domestic chores 家务杂活
prolific *a.* 作品多的，多产的
re-invigorate *v.* 再次激励
inter *v.* 埋葬
epitaph *n.* 碑文，墓志铭
gravesite *n.* 墓地，坟墓

Architectural Terms

civil engineering 土木工程（学）
residential *a.* 居住的，住所的
prairie house 草原住宅
crucifix *n.* 十字架（型）
plaster *n.* 灰泥，灰浆

Proper Names

Adler and Sullivan 阿德勒和沙利文（建筑事务所）
Prairie school 草原学派（19世纪末20世纪初以赖特为首的芝加哥年轻建筑师共同打造的学派，探究建筑的水平发展）
Arts and Crafts Movement 艺术与工艺运动（19世纪后期至20世纪早期兴起于英美的工艺美术运动）
Taliesin 塔里埃森
Imperial Hotel （东京）帝国饭店
Taliesin Fellowship 塔里埃森设计团体
Fallingwater 流水别墅
Johnson Administration Building 约翰逊（制蜡公司）办公大楼
Usonian house "桑年"房屋（赖特自创的具有美国本土特色的房屋）
Price Tower 普赖斯大厦
the Guggenheim Museum 古根海姆美术馆（纽约）
Marin County Civic Center 马林文娱中心
Taliesin West 西塔里埃森
Unity Chapel 唯一教堂

EXERCISES

I. Match the English expressions with their Chinese equivalents.

1. civil engineering	A. 十字架（型）
2. residential	B. 古根海姆美术馆
3. overlapping	C. 流水别墅
4. crucifix	D. 灰泥；灰浆
5. plaster	E. 制图员，绘图家
6. Prairie school	F. 土木工程（学）
7. Arts and Crafts Movement	G. 草原学派
8. the Guggenheim Museum	H. 重迭的；交搭的
9. draftsman	I. 居住的；住所的
10. Fallingwater	J. 艺术与工艺运动

II. Decide whether the following statements are True or False.

1. Frank Lloyd Wright's father was a teacher and his mother a musician and preacher.
2. Adler and Sullivan Firm was at the leading edge of American architecture at the time of the end of the 19th century.
3. Frank Lloyd Wright's "Prairie House" concept is a long, low building with hovering planes and vertical emphasis.
4. Few commissions were completed by Wright between 1914 and 1932, but he did lecture and publish frequently, with books including *An Autobiography* in 1932.
5. Wright never retired, and during the last part of his life, Wright produced a wide range of work including a retreat in Phoenix, Arizona.

III. Choose the best answer to each of the following questions.

1. Who was Wright's beloved teacher?
   A. J. Lyman Silsbee. B. Louis Sullivan. C. Dankmar Adler. D. Mamah Cheney.
2. When did Wright start his own firm?
   A. In 1867. B. In 1887. C. In 1893. D. In 1909.
3. Which one is regarded as the most famous work designed by Wright?
   A. The Imperial Hotel in Tokyo. B. Prairie House.
   C. The Usonian house. D. Fallingwater.
4. Where is Taliesin, Wright's studio, located?
   A. In Wisconsin. B. In Chicago. C. In Germany. D. In New York.

5. Which period was the most productive period for Wright?
   A. 1893-1909. B. 1914-1932. C. The war years. D. The post-war period.

IV. Oral task

In next class, you'll be asked to give an oral report based on one of the following questions. Work in teams and search the library or Internet for relevant pictures, facts or stories to support your points.

→ 1 What do you know about Wright's idea as opposed to the view of single-function-box rooms and enclosed space?

→ 2 What do you know about Wright's "Prairie House" concept?

→ 3 What do you know about Taliesin Fellowship?

TEXT B

I believe that architecture is a pragmatic art. To become art it must be built on a foundation of necessity. It is easy to say that the art of architecture is everything, but how difficult it is to introduce the conscious intervention of an artistic imagination without straying from the context of life. —Ieoh Ming Pei's Acceptance Speech of the Pritzker Architecture Prize

Ieoh Ming Pei

Ieoh Ming Pei was born in Canton, China in 1917, the son of a prominent banker. At age seventeen he went to the United States to study architecture, and received a Bachelor of Architecture degree from MIT in 1940. Upon graduation, he was awarded the Alpha Rho Chi Medal, the MIT Traveling Fellowship, and the AIA Gold Medal. In 1942, Pei enrolled in the Harvard Graduate School of Design, where he studied under Walter Gropius; six months later, he volunteered his services to the National Defense Research Committee in Princeton. Pei returned to Harvard in 1944 and completed his Master of Architecture in 1946, simultaneously teaching on the faculty as an assistant professor (1945-1948). Awarded the Wheelwright Traveling Fellowship by Harvard in 1951, he traveled extensively in England, France, Italy, and Greece. I. M. Pei became a naturalized citizen of the United States in 1954.

In 1948, William Zeckendorf invited Pei to accept the newly created post of Director of Architecture at Webb & Knapp, a real estate development corporation, resulting in many large-scale architectural and planning projects across the country. In 1955 he formed the partnership of I. M. Pei & Associates, which became I. M. Pei & Partners in 1966, and Pei Cobb Freed & Partners in 1989. The partnership received the 1968 Architectural Firm Award of the American Institute of Architects.

Mr. Pei's personal architectural style blossomed with his design for the National Center for Atmospheric Research in Boulder, Colorado (1961-1967). He subsequently gained broad national attention with the East Building of the National Gallery of Art in Washington D.C. (1968-1978) and the John F. Kennedy Library in Boston (1965-1979)—two of some thirty institutional projects executed by

Pei, including church, hospital, and municipal buildings, as well as schools, libraries, and over a dozen museums. His most recent works include the Miho Museum in Shiga, Japan, the Grand Louvre Pyramids in Paris, the Morton H. Meyerson Symphony Center in Dallas, and the Schauhaus at the German Historical Museum in Berlin. Among Pei's skyscraper designs are the 72-storey Bank of China in Hong Kong and the Four Seasons Hotel in midtown Manhattan. Other hotels by Pei include Raffles City in Singapore and Fragrant Hill Hotel in Beijing (1982), which was designed to graft advanced technology onto the roots of indigenous building and thereby sow the seed of a new, distinctly Chinese form of modern architecture.

▲ National Gallery of Art

Due to his reliance on abstract form and materials such as stone, concrete, glass, and steel, Pei has been considered a disciple of Walter Gropius. However, he shows little concern with theory. He does not believe that architecture must find forms to express the times or that it should remain isolated from commercial forces. Pei generally designs sophisticated glass-clad buildings loosely related to the high-tech movement. However, many of his designs result from original design concepts. He frequently works on a large scale and is renowned for his sharp, geometric designs.

Mr. Pei's deep interest in the arts and education is evidenced by his numerous memberships on Visiting Committees at Harvard and MIT, as well as on several governmental panels. He has also served on the AIA Task Force on the West Front of the U.S. Capitol. A member of the AIA National Urban Policy Task Force and of the Urban Design Council of the City of New York, he was appointed to the National Council on the Humanities by President Lyndon Johnson in 1966, and to the National Council on the Arts by President Jimmy Carter in 1980. In 1983, Mr. Pei was chosen the Laureate of the Pritzker Architecture Prize; he used the $100,000 award to establish a scholarship fund for Chinese students to study architecture in the United States (with the strict proviso that they return to China to practice their profession). Among the many academic awards bestowed on Mr. Pei are honorary doctorates from Harvard University, the University of Pennsylvania, Columbia University, New York University, Brown University, the University of Colorado, the Chinese University of Hong Kong, and the

American University of Paris. Most recently he was awarded the Laura Honoris Causa by the University of Rome, in 2004.

Ieoh Ming Pei has given this century some of its most beautiful interior spaces and exterior forms. Yet the significance of his work goes far beyond that. His concern has always been the surroundings in which his buildings rise.

VOCABULARY

New Words

naturalized *a.* 加入国籍的
municipal *a.* 市政的，市的
graft *v.* 移植，嫁接
disciple *n.* 信徒，门徒
proviso *n.* （附带）条件，条款
bestow *v.* 给予，赠与
honorary *a.* 荣誉的
doctorate *n.* 博士学位

Architectural Terms

real estate 房地产
glass-clad *a.* 玻璃覆盖的

Proper Names

Ieoh Ming Pei 贝聿铭
Canton 广州
MIT 麻省理工学院
AIA (the American Institute of Architects) 美国建筑师协会
William Zeckendorf 威廉·柴根道夫（美国房地产开发商）
Webb & Knapp 韦伯·纳普（建筑公司）
National Center for Atmospheric Research （美国）国家大气研究中心
National Gallery of Art （美国）国家美术馆
John F. Kennedy Library 肯尼迪总统图书馆
Miho Museum （日本）美秀博物馆
the Grand Louvre Pyramids （法国）卢浮宫金字塔
the Morton H. Meyerson Symphony Center 梅尔森音乐厅
Schauhaus （德国历史博物馆的）展馆；展室
the German Historical Museum 德国历史博物馆
Bank of China 中国银行
the Four Seasons Hotel 四季酒店
Raffles City 莱福士广场
Fragrant Hill Hotel （北京）香山饭店
Lyndon Johnson 林登·约翰逊（美国第36任总统）
Jimmy Carter 吉米·卡特（美国第39任总统）

EXERCISES

I. Translate the following English into Chinese and Chinese into English.

1. real estate ______________
2. the Grand Louvre Pyramids ______________
3. glass-clad ______________
4. geometric design ______________
5. indigenous ______________
6. 城市设计 ______________
7. （附带）条件；条款 ______________
8. 合作企业；合作关系 ______________
9. 市政大楼 ______________
10. 移植，嫁接 ______________

II. Reorganize the following statements about Ieoh Ming Pei according to the time sequence stated in the passage.

a. He returned to Harvard.
b. He received a Bachelor of Architecture degree from MIT.
c. He was born in Canton, China.
d. He became a naturalized citizen of the United States.
e. He went to the United States to study architecture.
f. He formed the partnership of I. M. Pei & Associates.
g. He was awarded the Alpha Rho Chi Medal, the MIT Traveling Fellowship, and the AIA Gold Medal.
h. He was awarded the Wheelwright Traveling Fellowship by Harvard.
i. He was appointed to the National Council on the Humanities by President Lyndon Johnson.
j. He enrolled in the Harvard Graduate School of Design.
k. He was invited by William Zeckendorf to accept the newly created post of Director of Architecture at Webb & Knapp.
l. He completed his Master of Architecture.
m. He was chosen the Laureate of the Pritzker Architecture Prize.
n. He was appointed to the National Council on the Arts by President Jimmy Carter.

1	2	3	4	5	6	7
□	□	□	□	□	□	□
8	9	10	11	12	13	14
□	□	□	□	□	□	□

IV. Oral task

In next class, you'll be asked to give an oral report based on one of the following questions. Work in teams and search the library or Internet for relevant pictures, facts or stories to support your points.

→ 1 Why has Pei been considered a disciple of Walter Gropius?

→ 2 What are the main characteristics of Pei's works on a large scale?

→ 3 Name one world-renowned architecture designed by Ieoh Ming Pei and tell the class your understanding of it.

TEXT C

Famous Ancient Architects

▲ Marcus Vitruvius Pollio

▲ image from *Ten Books on Architecture*

Marcus Vitruvius Pollio

Marcus Vitruvius Pollio (80/70-25 BC), Roman writer, architect, and engineer, is best known as the author of the handbook for Roman architects, *De architectura libri decem* (*Ten Books on Architecture*), which carefully described existing practices, not only in the design and construction of buildings, but also in what are today thought of as engineering disciplines. His books include such varied topics as the manufacture of building materials and dyes (material science), machines for heating water for public baths (chemical engineering), amplification in amphitheaters (acoustics), and the design of roads and bridges (civil engineering). His writing is prescriptive and gives direct advice: "I have drawn up definite rules to enable you, by observing them, to have personal knowledge of the quality both of existing buildings and of those which are yet to be constructed." As a handbook, *De architectura* was wildly successful, and Vitruvius' advice was followed for centuries.

Filippo Brunelleschi

Filippo Brunelleschi (1377-1446), Florentine architect, is one of the initiators of Italian Renaissance architecture. His revival of classical forms, and his championing of architecture based on mathematics, proportion, and perspective, makes him a key artistic figure in the transition from the Middle Ages (5th century to 15th century) to the modern era.

▲ the dome of the Cathedral in Florence

Brunelleschi was born in Florence and in 1418 was commissioned to build the dome of the unfinished Florence

▲ Filippo Brunelleschi

Cathedral. The dome, a great innovation both artistically and technically, is one of the first examples of architectural functionalism. Brunelleschi devised an austere, geometric style inspired by the art of ancient Rome and completely different from the emotional, elaborate Gothic mode that prevailed. His use of straight lines, flat planes, and cubic spaces influenced later buildings of the Florentine Renaissance. Later in Brunelleschi's career, he developed a somewhat more sculptural, rhythmic style that was the first step toward Baroque architecture. Brunelleschi influenced his contemporaries and immediate followers, as well as modern architects, who came to revere him as the first great exponent of rational architecture.

Michelangelo

▲ Michelangelo

Michelangelo (1475-1564), Italian artist, is one of the most inspired creators in the history of art. As a sculptor, architect, painter, and poet, he exerted a tremendous influence on his contemporaries and on subsequent Western art. Born in Caprese, Michelangelo is most associated with the city of Florence. His father, a Florentine official who had connections to the ruling Medici family, placed his 13-year-old son in the workshop of Florentine painter Domenico Ghirlandaio. Later, after studying at the sculpture school in the Medici gardens, Michelangelo was invited into the household of Lorenzo de' Medici.

Michelangelo's activity as an architect began in 1519, but his plan for the facade of Florence's Church of San Lorenzo was never executed. In the 1520s he designed the Laurentian Library and its elegant entrance hall adjoining San Lorenzo, although these were not finished until decades later. Between 1519 and 1534 Michelangelo also undertook the commission of the Medici Tombs for the New Sacristy of San Lorenzo. Work on the tombs continued long after Michelangelo went back to Rome in 1534.

From 1536 to 1541 Michelangelo worked on the altar wall of the Sistine Chapel, creating Last Judgment, the largest fresco of the

▼ Michelangelo Plaza

Renaissance. He was also commissioned to paint the Pauline Chapel in the 1540s, but he directed his main energies toward architecture. Although his program was not finished until the 17th century, Michelangelo designed the remodeling of the buildings surrounding the Campidoglio on the Capitoline Hill (Monte Capitoline), the civic and political heart of Rome. His crowning achievement as an architect was his work at Saint Peter's Basilica, where he was made chief architect in 1546. The building was constructed according to plans by Italian architect Donato Bramante, but Michelangelo ultimately became responsible for the exterior altar end of the building and for the dome's final form.

▲ Sir Christopher Wren

▲ St. Paul's Cathedral

Sir Christopher Wren

Christopher Wren (1632-1723) was born in Wiltshire, England. He attended Wadham College, Oxford in 1649. At Oxford he joined a group of brilliant scholars, who later formed the core of the Royal Society. As assistant to an eminent anatomist, Wren developed skills as an experimental, scientific thinker. With astronomy as his initial course of study, Wren developed skills in working models, diagrams and charting that proved useful when he entered architecture. In 1663, Wren's uncle, the Bishop of Ely, asked him to design a new chapel for Pembroke College, Cambridge. This, his first foray into architecture, was quickly followed by more commissions. London's Great Fire of 1666 gave Wren a chance to present a scheme to rebuild the city. Utopian in concept, it was only partially realized. In 1669 Charles II appointed Wren Surveyor General of the King's Works. As Surveyor General he supervised all work on the royal palaces. In 1673 Wren resigned his Oxford professorship because of the workload. He was also knighted in 1673.

VOCABULARY

New Words

amplification *n.* 扩音
amphitheater *n.* （古罗马）圆形的露天剧场
acoustics *n.* （房间，剧场等的）音质，声学
prescriptive *a.* 规定的，指示的
initiator *n.* 创始人
austere *a.* 朴素的，无装饰的
revere *v.* 尊敬，崇敬
adjoin *v.* 贴近，邻接
civic *a.* 城市的，市民的
anatomist *n.* 解剖学家
astronomy *n.* 天文学
bishop *n.* 主教
foray *n.* 突袭
Utopian *a.* 乌托邦的
knight *v.* 封为爵士（或骑士）

Architectural Terms

functionalism *n.* 功能主义建筑（主张把建筑的实用功能放在设计的首位）
sculptural *a.* 雕刻的，雕塑的
altar *n.* 祭坛，祭台

Proper Names

Florentine 佛罗伦萨（人）的；佛罗伦萨文化的
Domenico Ghirlandaio 多米尼哥·吉兰达约（1449–1494，文艺复兴初期佛罗伦萨画家，曾为梵蒂冈西斯廷礼拜堂作画）
Sistine Chapel 西斯廷礼拜堂（梵蒂冈的主要教堂，以米开朗琪罗及其他艺术家的天顶画和壁画著称）

Answer the following questions according to the text.

1. What do the books *De architectura libri decem* mainly include?
2. What characteristics of Brunelleschi make him a key artistic figure in the transition from the Middle Ages to the modern era?
3. Who is regarded as the first great exponent of rational architecture?
4. What was Michelangelo's highest achievement as an architect?
5. What can we know about Sir Christopher Wren from the text?

World-Famous Buildings

Warming Up

1. There are many world-famous buildings. Can you list some of them and say what their names are, where they locate, when they were built?
2. Do you know any interesting stories about the famous buildings in your country?

TEXT A

The Greatest Architecture of the Past 1,000 Years

What are the most significant, most beautiful, or most interesting buildings of the past 1,000 years? Some art historians choose the Taj Mahal, while others prefer the soaring skyscrapers of the 20th century. There is no single correct answer. Perhaps the most innovative buildings are not grand monuments, but obscure homes and temples. In this quick list, we'll take a whirlwind tour through time, visiting some of the most popular buildings and a few forgotten treasures.

Church of St. Denis

1. St. Denis Church in Saint-Denis (1137?)

During the Middle Ages, builders were discovering that stone could carry far greater weight than ever imagined. Cathedrals could soar to dazzling heights, yet create the illusion of lace-like delicacy. The Church of St. Denis, commissioned by Abbot Suger of St. Denis, was one of the first large buildings to use a new vertical style known as Gothic. The church became a model for most of the late 12th century French cathedrals, including Chartres.

2. Chartres Cathedral Reconstruction (1205-1260)

In 1194, the original Romanesque Chartres Cathedral in Chartres, France was destroyed by fire. Reconstructed in the years 1205 to 1260, the new Chartres Cathedral was built in the new Gothic style. Innovations in the cathedral's construction set the standard for thirteenth-century architecture.

Chartres Cathedral

Layout of the Forbidden City

3. The Forbidden City, Beijing (1406-1420)

Occupying a rectangular area of more than 720,000 square meters, the Forbidden City was the imperial home of 24 emperors of the Ming (1368-1644) and Qing (1644-1911) dynasties. The Forbidden City is one of the largest and best-preserved palace complexes in the world. There are over a million rare and valuable objects in the Museum.

the Louvre

4. The Louvre, Paris (1546 and Later)

In the late 1500s, Pierre Lescot designed a new wing for the Louvre and popularized ideas of pure classical architecture in France. Lescot's design laid the foundation for the development of the Louvre over the next 300 years. In 1985, architect Ieoh Ming Pei stirred great controversy when he designed the stark glass pyramid entrance to the palace-turned-museum.

5. Palladio's Basilica, Italy (1549 and Later)

During the late 1500s, Italian Renaissance architect Andrea Palladio brought new appreciation for the classical ideas of ancient Rome when he transformed the town hall in Vicenza, Italy into the Basilica (Palace of Justice). Palladio gave the remodeled building two styles of classical columns: Doric on the lower portion and Ionic on the upper portion. Palladio's later designs continued to reflect the humanist values of the Renaissance period.

Palladio's Basilica

Taj Mahal

6. Taj Mahal, India (1630-1648)

According to legend, the Mughal emperor Shah Jahan wanted to build the most beautiful mausoleum on Earth to express his love for his favorite wife. Or, perhaps he was simply asserting his political power. The Taj Mahal may have been designed by Ustad Ahmad Lahori, an Indian architect of Persian descent. Persian, Central Asian, and Islamic elements combine in the great white marble tomb. The Taj Mahal is just one of many architectural wonders in a land of majestic tombs and erotic temples.

7. Monticello, Virginia, USA (1768-1782)

When the American statesman, Thomas Jefferson, designed his Virginia home, he combined the European traditions of Palladio with American domesticity. Jefferson's plan for Monticello resembles Palladio's Villa Rotunda. With a few innovations, Jefferson gave Monticello long horizontal wings, underground service rooms, and "modern" conveniences.

Monticello in Virginia

the Eiffel Tower

8. The Eiffel Tower, Paris (1889)

The Industrial Revolution in Europe brought about a new trend: the use of metallurgy in construction. Because of this, the engineer's role became increasingly important, in some cases melding with or rivaling that of the architect. The Eiffel Tower is the tallest building in Paris, and reigned for 40 years as the tallest in the world.

9. The Wainwright Building, St. Louis, Missouri (1890)

Louis Sullivan and Dankmar Adler redefined American architecture with the Wainwright Building in St. Louis, Missouri. Their design emphasized the underlying structure. Except for the large, deep windows, the first two storeys are unornamented. Uninterrupted piers extend through the next seven stories. Intertwined ornaments and small round windows form the upper story. "Form follows function," Sullivan told the world.

the Wainwright Building

Sidney Opera House

10. Great Buildings of the 20th Century

During the 20th century, exciting new innovations in the world of architecture brought soaring skyscrapers and fresh new approaches to home design, to name just a few, the 77-storey Chrysler Building, the Empire State Building, and the 110-storey World Trade Center (known as the "Twin Towers" and destroyed in 2001) in New York, Sidney Opera House in Australia, and of course Frank Lloyd Wright's improbable structure—Fallingwater.

VOCABULARY

New Words

obscure *a.* 不引人注意的
whirlwind *a.* 旋风般的，快速的
dazzling *a.* 令人赞叹不已的
lace-like *a.* 像花边一样的
delicacy *n.* 优美，精致
stark *a.* 轮廓分明的，显眼的
majestic *a.* 庄严的，雄伟的
domesticity *n.* 本土（特点）
meld *v.* 混合
intertwined *a.* 缠绕在一起的

Architectural Terms

mausoleum *n.* 陵墓
metallurgy *n.* 冶金学，冶金术

Proper Names

St. Denis Church 圣丹尼斯教堂
Pierre Lescot 皮埃尔·莱斯科（法国建筑师）
Palladio's Basilica 帕拉迪诺教堂
Andrea Palladio 帕拉迪诺（1508–1570，意大利建筑师，常被认为是西方最具影响力的建筑师）
Vicenza 维琴察（意大利东北的一座城市）
the Wainwright Building 韦恩赖特大厦

EXERCISES

I. Match the English expressions with their Chinese equivalents.

1. metallurgy	A. 宫殿（建筑）群
2. Romanesque	B. 罗马风格（的）
3. rectangular	C. 波斯的
4. palace complex	D. 冶金术
5. wing	E. 陵墓
6. mausoleum	F. 文艺复兴
7. pier	G. 长方形，矩形
8. Renaissance	H. 紫禁城
9. Persian	I. 支柱，扶壁
10. the Forbidden City	J. 侧翼，厢房

II. Decide whether the following statements are True or False.

1. The Church of St. Denis, commissioned by Abbot Suger of St. Denis, was the first large building to use new vertical style known as Gothic.
2. The Forbidden City is one of the largest and best-preserved palace complexes in the world.
3. Thomas Jefferson's plan for Monticello resembles Palladio's Villa Rotunda without any innovations.
4. The Eiffel Tower is the tallest building in Paris, and reigned for almost 100 years as the tallest in the world.
5. "Form follows function" was Louis Sullivan's idea about architecture.

III. Choose the best answer to each of the following questions.

1. The original _____ Chartres Cathedral in Chartres was destroyed by fire and was reconstructed in the years 1205 to 1260. The new Chartres Cathedral was built in _____style.

   A. Gothic; Roman　　B. Gothic; Romanesque

   C. Roman; Gothic　　D. Romanesque; Gothic

2. In the late 1500s, Pierre Lescot designed a new _____ for the Louvre and popularized ideas of pure _____ architecture in France.

   A. wing; classical　　B. facade; classical

   C. wing; neoclassical　　D. facade; neoclassical

3. ________ architectural elements combine in the great white marble tomb, Taj Mahal.
   A. Indian, Central Asian, and Islamic B. Persian, Central Asian, and Islamic
   C. Indian, East Asian, and Islamic D. Persian, East Asian, and Islamic
4. The coming of ________ in Europe brought about a new trend: the use of ________ in construction. Therefore an engineer's role became increasingly important, in some cases equal to that of an architect.
   A. the Building revolution; metallurgy B. the Building revolution; machinery
   C. the Industrial Revolution; metallurgy D. the Industrial Revolution; machinery
5. By emphasizing the underlying structure, ________ firstly redefined American architecture with the Wainwright Building in St. Louis, Missouri.
   A. Louis Sullivan and Lloyd Wright B. Louis Sullivan and Dankmar Adler
   C. Lloyd Wright and Dankmar Adler D. Lloyd Wright and Thomas Jefferson

IV. Oral task

In next class, you'll be asked to give an oral report based on one of the following questions. Work in teams and search the library or Internet for relevant pictures, facts or stories to support your points.

→ 1 Suppose you are asked to be a tour guide for a pen-friend who wants to visit the Forbidden City, what information will you include in your introduction?

→ 2 Search any interesting information about the Taj Mahal and share it with your classmates.

→ 3 Why did Emperor Shah Jahan want to build the most beautiful mausoleum on Earth?

TEXT B

Fallingwater

Fallingwater is a luxury vacation retreat Frank Lloyd Wright designed in 1936 for Bear Run, Pennsylvania, and Pittsburgh millionaire Edgar Kaufmann. Cantilevered over a waterfall, this magnificent home has probably been photographed, written about, analyzed, and applauded more than any other Wright building. Not only did it win him enormous acclaim at the time but it is regarded as one of his finest structures ever built. One critic calls it "the most famous modern house in the world", while another says it is "one of the complete masterpieces of 20th-century art". The architect exploited his opportunity to the fullest, producing his most impressive residence ever, one that hundreds of thousands of people have gone out of their way to see and that almost everyone acknowledges to be virtually unsurpassed.

One of the most remarkable characteristics of Fallingwater is its absolute refusal to be confined. Fallingwater seems to take flight every way at once, making it exceptionally difficult to visualize or to describe to someone who has not seen it. (Most photographs do not completely capture it.) This, in fact, may have been one of Wright's objectives: to defy description, to destroy categories. "It has no limitations as to form," he once remarked. So difficult visually to comprehend—so impossible to harness, as it were—Fallingwater disrupted expectations about what a house should be or do. Visitors are surprised, for example, by its comparatively few rooms, assuming it to be much larger than it is. Most of its floor space, surface area, and expense were devoted to a massive living and dining room and to terraces and canopy slabs shooting out in several directions, while its three bedrooms and the usual services take up a small proportion of its three levels. Fallingwater was not so much a family residence as a weekend entertainment retreat and, like the Guggenheim Museum in New York, is partly an exercise in architectural sculpture. It should not be accused of "impracticality" or unnecessary expense; for as Henry-Russel Hitchcock has wisely stated, architecture lives not only "through the solution of generic problems [to which Wright devoted

considerable attention] but quite as much by the thrill and acclaim of unique masterpieces".

Like few other buildings before or since, Fallingwater exploited site to advantage. Two unbelievably cantilevered terraces, partially sheltered one by the other and by slab roof canopies, cross twelve feet above a waterfall in Bear Run which passes in front of the structure's ledge on which it rests, while the vertical trust of a stone fireplace and chimney stakes the house firmly in place, echoing the plunging stream, reaching for the sky. Its composition is an abstract reformulation of its natural site, a poetic but not a literal interpretation of the defining features of its locale. Without Fallingwater, Bear Run would have remained a beautiful forest stream, like thousands of others; but with it, the place became unique. Here was an unsurpassed example of humanity making the world a better place, of art working to improve nature. Fallingwater also supported Wright's contention that an organic building was appropriate only on its particular spot, and nowhere else.

But Fallingwater achieved its truest measure of greatness in the way it transcended site to speak to universal human concerns. In its startling departure from traditional modes of expression, it revealed an aspiration for freedom from imposed limitations; and in its successful partnership with the environment, it was a guidepost to humanity's proper relationship with nature. Fallingwater was also a resolution of dichotomies. At the same time strikingly substantial and dangerously ephemeral, it is securely anchored to rock and ledge, but seems to leap into space. It embodies change and changelessness simultaneously, for its imperishable stone and concrete elements from entirely new compositions as the angle of vision shifts. Solid rock and rushing water reflect the permanence and impermanence of life itself. Fallingwater sinks its roots deeply into the ground to grow out of its site more like a plant than most other buildings, yet it is a masterpiece of sophisticated construction techniques. Composed of lots of rectangles, it is never redundant; built of innumerable pieces of varying size and material, it nevertheless achieves a unity few structures approach. Fallingwater is a study in opposites—motion and stability, change and permanence, power and ephemeralness—that make the human condition a paradox of welcome adventure and anxious uncertainty. The philosophically ambiguous house at Bear Run may have been a comment on the social contradiction of a rich man's wealth in times of general depression, but it was also Wright's nature poem to modern humanity.

VOCABULARY

New Words

acclaim *n.* 欢呼，称赞
unsurpassed *a.* 无法超越的
exceptionally *adv.* 非凡地，尤其地
visualize *v.* 想象，设想
generic *a.* 普遍的，类属的
contention *n.* 论点
transcend *v.* 超越
guidepost *n.* 指示牌，路标
dichotomy *n.* 一分为二（尤指成对立的两部分）
ephemeral *a.* 短暂的
imperishable *a.* 不易腐坏的
paradox *n.* 自相矛盾的事物
philosophically *adv.* 哲学地

Architectural Terms

cantilever *v.* 如悬臂向外伸展
canopy *n.* 华盖，天篷
slab *n.* 平板，平板状物
ledge *n.* 壁架，窗台
locale *n.* 场所，地点

Proper Names

Henry-Russel Hitchcock 亨利·拉塞尔·希契科克（建筑史学家）

EXERCISES

I. Translate the following English into Chinese.

1. cantilever ______________
2. terrace ______________
3. canopy ______________
4. slab ______________
5. ledge ______________
6. locale ______________
7. guidepost ______________
8. dichotomy ______________
9. organic ______________
10. residence ______________

II. Complete the table according to the given information.

Features of Fallingwater	Further Descriptions of Fallingwater
its absolute refusal to be confined	1. Fallingwater ________ expectations about what a house should be or do. 2. Fallingwater is partly an exercise in architectural ________.
its exploiting site to advantage	3. Its composition is an abstract reformulation of its ________. 4. Two unbelievably ________ terraces, partially sheltered one by the other and by slab roof canopies.
its truest measure of greatness	5. It transcended site to speak to universal ________ concerns. 6. It embodies motion and ________, change and ________, power and ________—that make the human condition a(n) ________ of welcome adventure and anxious uncertainty.

IV. Oral task

In next class, you'll be asked to give an oral report based on one of the following questions. Work in teams and search the library or Internet for relevant pictures, facts or stories to support your points.

→ 1 Explain and illustrate your understanding of Wright's contention that an organic building was appropriate only on its particular spot.

→ 2 In what way can we see that Fallingwater is a study in opposites?

TEXT C

Of all the Grand Projects in Paris, none created such a stir as the Pei Pyramids in the courtyard of the famous Louvre Museum.

— Dennis Sharp, *Twentieth Century Architecture: A Visual History*

The Grand Louvre Pyramids

Constructed of cut stone, the Louvre is a masterpiece of the French Renaissance. Architect Pierre Lescot was one of the first to apply pure classical ideas in France, and his design for a new wing at the Louvre defined its future development.

With each new addition, under each new ruler, the palace-turned-museum continued to make history. Its distinctive double-pitched mansard roof inspired the design of many 18th-century buildings in Paris and throughout Europe and the United States. The Louvre is the most famous of the Grand Projects—Mitterrand's 15 billion franc program to provide a series of modern monuments to symbolize France's central role in art, politics, and world economy at the end of the 20th century.

Traditionalists were shocked when Chinese-born American architect I. M. Pei designed three glass pyramids at the entrance to the Louvre in 1985. Pei's design consisted of unusual arrangements of geometric shapes. In order to preserve the views of the historical facades and to harmonize with the classical geometry and symmetry, this entrance structure was chosen to be a centrally located glass pyramid. The steel frame of this entrance is 116 feet on each side and 71 feet tall, giving the shallow "Hall Napoleon" entrance room below a strong presence above ground and serving as an immense skylight of well over 10,000 square feet in area. The spiderweb of steel members support 603 diamond shaped and 70 triangular panes of 21 millimeter thick glass, giving the structure a very elegant and transparent appearance. It was Pei's intention to utilize a large number of small members rather than a few larger ones to allow a better transparency. The diagonal panes have diagonals of 9.84 feet and 6.23 feet and the triangular panes are half of this size. The structure rests on four massive posts of reinforced concrete and steel which extend over two floors into the foundation below.

The main Pyramid is basically a complex inter-linked steel structure covered in reflective glass. In fact it is an entrance doorway providing a long entrance portico to the main galleries of the Louvre. As one descends into the interior entrance foyer, the dramatic nature of the intervention becomes apparent. The main Pyramid, which certainly disturbs the balance of the old Louvre courtyard, is countered by two smaller pyramids, which provide further light and ventilation to the subterranean spaces.

▲ the glass-covered steel structure

It is not so much the pyramid, but the entrance space that it covers that is the most important part of the project. The brilliance of making an entrance to the world's largest art museum by hollowing-out its plaza and constructing underground connections to its various wings could easily be lost amidst the unmistakable iconography of the pyramid. The entrance has rationalized and opened up the collections of the Louvre to the throngs of museum-goers who visit its collections.

Throughout, the restrained detailing of the stone walls and floors, simple geometries, and generously proportioned spaces serve as an appropriate backdrop against both new and old. The intricate steel connections and rods that support the pyramid are a 20th-century expression equivalent to the ornate carving in the masonry facades of the Palace of Louvre. The project also included the construction of a shopping mall, a cultural center, an auditorium, and parking garages. Although Pei's design was initially very controversial, anyone who visited the Louvre prior to this new entrance can recall standing in long lines and walking forever once inside (inevitably getting lost!). Without changing any of the exterior architecture, Pei created this entrance by excavating the Napoleon Court, formerly a parking lot in the center of the wings. The lower level, then, serves not only as the infrastructure for the museum but provides easy and clear access to the three wings of the museum.

After Pei's former firm, Pei Cobb Freed, completed the 70-foot-tall Pyramids of Louvre and 665,000-square-foot underground entry and concourse in 1989, Pei and museum officials agreed to move forward. Pei has confirmed that he is developing a solution to manage overcrowding in the Louvre's entrance. Pei says attendance at the Louvre has increased almost 70 percent since the popular project was completed, and now hosts 7.5 million visitors each year. Consequently, the public space has become crowded and loud, and has lost much of the peaceful aura the architect originally intended. "It's a real concern for me," says Pei, who had anticipated a less dramatic jump in attendance after his "Grand Louvre" project was completed. "If we don't do it, the place is going to look like an airport." Pei does not yet know the specific measures his team will carry out, but he says that a resolution should be sympathetic to the surrounding architecture. He adds that it should also allow visitors to access the collections despite construction.

Whether people love or hate the pyramid which sits at the center of the Louvre's plaza, the project—as large in scope as the image of the pyramid is well known—ultimately has to be appreciated at least for what it has accomplished in practical terms.

VOCABULARY

New Words

traditionalist *n.* 传统主义者，因循守旧的人
spiderweb *n.* 蜘蛛网
millimeter *n.* 毫米
amidst *prep.* (=amid) 在……当中，在……中间
rationalize *v.* 使合理化，合理地说明
throng *n.* 人群；大量；众多
backdrop *n.* 幕布，背景
intricate *a.* 错综复杂的，精心制作的
ornate *a.* 装饰华丽的
aura *n.* 气味，气氛，氛围

Architectural Terms

mansard roof 复折式屋顶，折线型屋顶
pane *n.* 长方块，长方格，窗格
transparency *n.* 透明性，透明度
diagonal *a.* 对角的，斜的
n. 斜构件，斜撑
reflective *a.* 反射的
foyer *n.* 门厅
ventilation *n.* 通风，空气的流通
subterranean *a.* 地下的
iconography *n.* 平面图
carving *n.* 雕刻

Proper Names

Mitterrand 密特朗（法国前总统）

Answer the following questions according to the text.

1. What was the purpose of Mitterrand's 15-billion-franc program of the Grand Projects?
2. Why did Pei design a centrally located glass pyramid as the entrance structure of the Louvre?
3. Why were two smaller pyramids designed with the main Pyramid in the old Louvre courtyard?
4. In addition to the Pyramids, what other constructions were also included in Pei's project?
5. What are the two main functions of the lower level of the Grand Louvre?

6
Unit

Chinese Architecture

Warming Up

1. How much do you know about Chinese architecture?
2. Compare an ancient Chinese building with a modern one. What are the differences between them as far as the architectural style is concerned?

TEXT A

On Chinese Architecture

Chinese architecture refers to a style of architecture that has taken shape in Asia over the centuries. The structural principles of Chinese architecture have remained largely unchanged, the main changes being only the decorative details. Since the Tang Dynasty, Chinese architecture has had a major influence on the architectural styles of Korea, Japan and Vietnam.

Chinese architecture enjoys a long history and great achievements, and has created many architectural miracles such as the Great Wall and the Forbidden City. In the process of its development, superior architectural techniques and artistic design were combined to make Chinese architecture a unique architectural system.

Basic Idea

Chinese architecture is based on the principle of balance and symmetry. Office buildings, residences, temples, and palaces all follow the principle that the main structure is the axis. The secondary structures are positioned as two wings on either side to form the main court and yard. The distribution of interior space reflects Chinese social and ethnical values. For example, a traditional residential building assigns family members based on the family's hierarchy.

Structure

As early as the Neolithic period, a basic principle of Chinese architecture was already established, wherein columns spaced at intervals, rather than walls, provided the support for the roof. Walls came to serve merely as enclosing screens. Chinese architecture features unique timber framework that clearly identifies supporting structure and bonding structure. The top load of a structure will be transferred to its foundations through its posts, beams, lintels and joists. Walls bear no loading and separate space only so that windows and walls will not be restricted to certain locations on the walls.

Color

Timber framework decides that color is the main ornament used on ancient Chinese architecture. In the beginning, paint was used on wood for antisepsis while later painting became an architectural ornament. In the feudal society, the use of color was restricted according to strict social status classification. Since yellow was deemed the noblest color and green and red the second, they were often applied on palace paintings. Usually, dragon or phoenix was painted on green background with mass gold powder or gold foil. The painting will give the structure a magnificent noble image under the background of white granite basement. It is unique that such sharp color can achieve artistic effects.

Roof

Although the typical Chinese roof was probably developed in the Shang Dynasty (1523-1027 BC) or the Zhou Dynasty (1027-256 BC), its features were unknown to us until the Han Dynasty. Then it appeared in the form that we recognize today as a hallmark of Chinese architecture—a graceful, overhanging roof, sometimes in several tiers, with upturned eaves. The roof rests on a series of four-part brackets (known as *toukong*), which in turn are supported by other clusters of brackets set on columns. The tremendous weight could eventually be brought down via the brackets system to the columns. Decorative possibilities were soon realized in the colorful glazed tiling of roofs and the carving and painting of brackets, which became more and more elaborate.

Roof is of great importance in Chinese architecture. It not only protects residents from the elements, but also has a deeper meaning. It either displays the honor of the owner of the house with its out-reaching upturned eaves, or for example, in a Buddhist temple, helps ward off evil spirits. The arc at which the roof turns comes from the intricate fit of rafters. Almost every wealthy home had elaborate roof on their house. One perfect example of splendid roof would be found inside the wonderful palaces, the Forbidden City and the Ming Tombs. They all have roof tiles which are brilliant yellow, green and red. The ridges of each roof carry figurines of mythical creatures. The curve of each roof can be no more than a sweep and the most intricate designs on the roof are almost always pointing southeast.

Ground Plan of Building Complex

During the Han Dynasty a characteristic ground plan was developed; it remained relatively constant through the centuries, applied to palaces and temple buildings in both China and Japan. Surrounded by an exterior wall, the building complex was arranged along a central axis and was approached by an entrance gate and then a spirit gate. Behind them in sequence came a public hall and finally the private quarters. Each residential unit was built around a central court with a garden. Based on imperial zoos and parks, the private residential garden soon became a distinctive feature of the walled complex and an art form in itself. The garden was laid out in a definite scheme, with a rest area and pavilions, ponds, and semi-planned vegetation.

Fengshui

Fengshui is a special Chinese tradition in architecture, and usually links the whole process from site selection, designing, construction and interior and exterior decorating in ancient times. *Feng* stands for wind or air and *shui* means water. *Fengshui* combines the trinity of the Heaven, the Earth and humans, and seeks harmony between selected site, orienting, natural doctrine and human fate. It repulses human destruction of nature and stresses cohabitation with the environment, which is regarded as perfect and occult.

VOCABULARY

New Words

axis *n.* 轴，轴线
position *v.* 安放，放置
ethnical *a.* 种族的
Neolithic *a.* 新石器时代的
antisepsis *n.* 防腐（法），抗菌（法）
feudal *a.* 封建（制度）的
deem *v.* 认为
phoenix *n.* 凤凰
foil *n.* 箔，金属薄片
hallmark *n.* 标志，特点
cluster *n.* 串，丛，簇
Buddhist *a.* 佛教的
ward off 避开，挡开
mythical *a.* 神话的
vegetation *n.* 植被
trinity *n.* 三位一体，三合一
repulse *v.* 击退，驱逐
cohabitation *n.* 共生，共同存在
occult *a.* 超自然的，神秘的

Architectural Terms

bonding *n.* 接，接合
joist *n.* 托梁
overhanging *a.* 突出的
tier building 多层建筑
bracket *n.* 托架，支架，斗拱
glazed tiling 釉面砖，琉璃瓦
rafter *n.* 椽
figurine *n.* 小雕像
spirit gate 神门

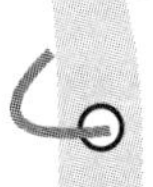

Proper Names

Tang Dynasty 唐朝
Vietnam 越南
Shang Dynasty 商朝
Zhou Dynasty 周朝
Han Dynasty 汉朝

EXERCISES

I. Match the English expressions with their Chinese equivalents.

1. timber	A. 椽
2. framework	B. 建筑群
3. lintel	C. 小雕像
4. granite	D. 木材，木料
5. upturned eaves	E. 托架，支架，斗拱
6. bracket	F. 过梁，楣
7. glazed tiling	G. 挑檐
8. rafter	H. 釉面砖，琉璃瓦
9. building complex	I. 花岗岩
10. figurine	J. 框架，结构

II. Decide whether the following statements are True or False.

1. Since the Shang Dynasty, Chinese architecture has had a major influence on the architectural styles of Korea, Japan, Vietnam, Malaysia, and Thailand.
2. The structural principles of Chinese architecture have remained largely unchanged since the ancient times, the main changes being only the decorative details.
3. Chinese architecture follows the principle that the main structures are positioned as two wings on either side and the secondary structures form the axis.
4. In Chinese architecture, the top load of a structure will be transferred to its foundations through its posts, beams, lintels and joists. Walls bear a part of the loading.
5. In Chinese architecture, the tremendous weight of roof could eventually be brought down via the brackets system to the columns.
6. Roof is of great importance in Chinese architecture because it not only protects residents from the elements, but also has a deeper meaning in that it can display the honor of its owner.

III. Choose the best answer to each of the following questions.

1. Chinese architecture refers to a style of architecture that has taken shape in ________.
   A. Japan B. Korea C. Asia D. Eastern Europe
2. The distribution of interior space of Chinese architecture reflects Chinese social and ________ values.
   A. economic B. political C. military D. ethnical
3. Chinese architecture features unique ________ framework that clearly identifies supporting structure and bonding structure.
   A. wooden B. stone C. concrete D. metal
4. Timber framework decides that ________ is the main ornament used on ancient Chinese architecture.
   A. shape B. color C. material D. form
5. A graceful ________ roof, sometimes in several tiers, with ________, is the most characteristic hallmark of Chinese architecture.
   A. hanging; upturned eaves
   B. overhanging; turned eaves
   C. hanging; turned eaves
   D. overhanging; upturned eaves
6. Surrounded by an exterior wall, the building complex was arranged along a central axis and was approached by ________. Behind them in sequence came ________.
   A. a spirit gate and then an entrance gate; a public hall and finally the private quarters
   B. a spirit gate and then an entrance gate; the private quarters and finally a public hall
   C. an entrance gate and then a spirit gate; a public hall and finally the private quarters
   D. an entrance gate and then a spirit gate; the private quarters and finally a public hall

IV. Oral task

In next class, you'll be asked to give an oral report based on one of the following questions. Work in teams and search the library or Internet for relevant pictures, facts or stories to support your points.

→ 1 What do you know about the principles that Chinese architecture follows? Illustrate your points by examples.

→ 2 How do you understand the supporting structure of Chinese architecture?

→ 3 How do you understand the Ming Tombs, considering its site selection, designing, construction, decoration, etc.?

TEXT B

Major Forms of Chinese Architecture

Chinese architecture is unique in the world of architectural styles. In terms of the form or formation and functions, it can be classified into six major types.

Gong (Palace)

After the founding of the Qin Dynasty, *gong* came to mean a group of buildings in which the emperors lived and worked. At about the same time, Chinese palaces grew ever larger in scale. The Efanggong of the First Emperor of Qin measured 2.5km from east to west and 1,000 paces from north to south. The Weiyanggong of the Western Han Dynasty had, within a periphery of 11 kilometers, as many as 43 halls and terraces. The Forbidden City of Beijing, which still stands intact and which served as the imperial palace for both Ming and Qing emperors covers an area of 720,000 square meters and embraces many halls, towers, pavilions and studies. It is one of the greatest palaces in the world. In short, the palace grew into a veritable city and is often called *gongcheng* (palace city). Apart from the palace, other grand buildings with similar scale and size are also called *gong* either for secular or religious use.

Ting (Pavilion)

A common sight in the country, a Chinese pavilion (*ting*) is built normally either of wood or stone or bamboo and may be in any of the shapes—square, triangle, hexagon, octagon, a five-petal flower, a fan and others. But all pavilions have columns for support and no walls. In parks or scenic areas, pavilions are built to command a panoramic view or built by the waterfront to create intriguing images. Pavilions serve diverse purposes. The wayside pavilion is

called *liangting* (cooling kiosk) to provide weary wayfarers with a place for rest. The "stele pavilion" gives a roof to a stone tablet to protect the engraved record of an important event. Pavilions also stand by bridges or over water-wells. In the latter case, dormer windows are built to allow the sun to cast its rays into the well as it has been the belief that water untouched by the sun would cause disease.

Tai (Building with Terrace)

The *tai* was an ancient architectural structure, a very much elevated terrace with a flat top. Generally built of earth, stone and surfaced with brick, they are used as a belvedere on which to look into the distance. In fact, however, many well-known ancient *tai* are not just a bare platform but has some palatial halls built on top.

Tai could be built to serve different practical purposes. For example, it could be used as an observatory as is the one near Jianguomen in Beijing which dates back to the Ming and Qing dynasties. It could also be used for military purposes like the beacon towers along the Great Wall, to transmit urgent information with smoke by day and fire by night. Also on the Great Wall, there is a square *tai* at intervals of every 300 to 400 meters from which the garrison troops kept watch. On the track of the ancient Silk Road, ruins of the old fortifications in the form of earthen terraces can still be seen.

Lou (Storeyed House)

Chinese character *lou* refers to any building of two or more storeys with a horizontal main ridge. The erection of such buildings began a long time ago in the Warring States Period, when *chonglou* (multi-storey house) was mentioned in historical records.

Ancient houses with more than one storey were meant for a variety of uses. The smaller two-storey house of private home generally has the owner's study or bedroom upstairs. The more magnificent ones built in parks or at scenic spots were belvederes from which to enjoy the distant scenery. In this case, it is sometimes translated as a "tower".

Ancient cities had bell and drum towers (*zhonglou* and *gulou*), usually palatial buildings with four-sloped, double-caved, glazed roofs, all-around verandas and colored and carved brackets supporting the overhanging eaves. Each tower housed a big bell or drum which was used to announce the time.

Ge (Storeyed Pavilion)

The Chinese character *ge* is similar to *lou* in that both are of two or more storey buildings. But *lou* has a door and windows only on the front side with the other three sides being solid walls. *Ge* is usually enclosed by wooden balustrades or decorated with boards all around.

Such buildings were used in ancient times for the storage of important articles and documents. Wenyuange for instance, in the Forbidden City of Beijing, was in effect the imperial library. Kuiwenge in the Confucius Temple of Qufu, Shandong Province was devoted to the safekeeping of the books and works of painting and calligraphy bestowed by the courts of various dynasties. Visitors to the city of Ningbo, Zhejiang Province, can still see Tianyige, which houses the greatest private collection of books handed down from the past. Monasteries of a large size normally have their own libraries built in the style of a *ge* and called Cangjingge to keep their collections of Buddhist scriptures. Some of the *ge*, notably those erected in parks, like other pavilions or towers (*ting*, *tai* and *lou*), were used for enjoying the sights.

Ta (Pagoda)

Derived from Indian stupa, *ta* or pagoda is a high rising structure which houses Buddha's relics. It is a combination of Buddhist philosophy and Chinese tradition. In the Tang Dynasty pagodas were usually simple, square structures; they later became more elaborate in shape and adornment.

In the 11th century a distinctive type of pagoda was created in the Liao territory. Built in three different stages, with a base, a shaft, and a crown, the structure was surmounted by a spire. Its plan was often octagonal, possibly as a result of the influence of Tantric Buddhism in which the cosmological scheme was arranged into eight compass points rather than four.

VOCABULARY

New Words

embrace *v.* 包含
veritable *a.* 真正的，名副其实的
secular *a.* 世俗的
hexagon *n.* 六角形，六边形
octagon *n.* 八角形，八边形
petal *n.* 花瓣
panorama *n.* 全景
waterfront *n.* 滨水地区
intriguing *a.* 引起兴趣的
diverse *a.* 不同的
wayfarer *n.* 旅客，徒步旅行者
palatial *a.* 宏伟的，壮丽的
beacon *n.* 灯塔，烽火信号
garrison troop 卫戍部队，驻军
fortification *n.* 防御工事，要塞
scripture *n.* 经文
relic *n.* 舍利子，遗物
adornment *n.* 装饰，装饰物
surmount *v.* 覆盖在……顶上
Tantric *a.*（佛教）密教哲学的
cosmological *a.* 宇宙哲学的，宇宙论的

Architectural Terms

kiosk *n.* 凉亭，小亭子
stele *n.* 石碑
tablet *n.* 牌匾，匾额
dormer window 天窗，老虎窗
belvedere *n.* 观景楼
veranda *n.* 游廊，走廊，阳台
pagoda *n.* 塔，宝塔
stupa *n.*（印度）佛塔，浮屠塔
spire *n.* 尖顶，塔尖

Proper Names

Qin Dynasty 秦朝
Efanggong 阿房宫
Weiyanggong 未央宫
Jianguomen 建国门（北京）
the Silk Road 丝绸之路
Warring States Period 战国时代
Wenyuange 文渊阁
Kuiwenge 魁文阁
Confucius Temple 孔庙
Qufu 曲阜（山东）
Ningbo 宁波（浙江）
Tianyige 天一阁
Cangjingge 藏经阁

EXERCISES

I. Translate the following English into Chinese and Chinese into English.

1. kiosk ______________
2. stele ______________
3. tablet ______________
4. dormer window ______________
5. belvedere ______________
6. 屋脊 ______________
7. 游廊，阳台 ______________
8. 栏杆，栅栏 ______________
9. 浮屠塔 ______________
10. 尖顶，塔尖 ______________

II. Decide whether the following statements are True or False.

1. After the founding of the Qin Dynasty, *gong* came to mean a group of buildings in which the emperors lived and worked. At about the same time, Chinese palaces grew ever larger in scale.
2. A Chinese pavilions built normally either of wood or stone or bamboo and only in a unique shape in plan—square.
3. The *tai* was an ancient architectural structure, a very much elevated terrace with a flat top. Generally built of earth, stone and surfaced with brick, they are used as a belvedere on which to enjoy the distant scene.
4. Chinese character *lou* refers to any building of one storey with a horizontal main ridge.
5. Such buildings called *ge* were usually used in ancient times for the storage of important articles and documents.
6. Derived from Indian stupa, *ta* or pagoda is a high rising structure which serves the living space for Buddhist monks.

III. Choose the best answer to each of the following questions.

1. The Forbidden City of Beijing, which still stands intact and which served as the imperial palace for both ______ emperors, covers an area of 720,000 square meters.
   A. Ming and Qing B. Song and Ming C. Tang and Song D. Han and Tang
2. In some of the pavilions, _______ are built to allow the sun to cast its rays into the well as it has been the belief that water untouched by the sun would cause disease.
   A. glass windows B. dormer windows C. door-like windows D. large windows

3. Ancient houses with more than one storey are meant for a variety of uses. The smaller two-storey house of private home generally has the owner's ______ upstairs.
   A. hall or dining B. hall or study C. study or dining D. study or bedroom
4. Ge is usually enclosed by ______ instead of ______ or decorated with boards all around.
   A. solid walls; wooden balustrades
   B. solid balustrades; wooden walls
   C. wooden walls; solid balustrades
   D. wooden balustrades; solid walls
5. *Ta* or pagoda is a combination of ______ philosophy and ______ tradition.
   A. Taoist; Chinese B. Buddhist; Chinese
   C. Chinese; Taoist D. Chinese; Buddhist

IV. Oral task

In next class, you'll be asked to give an oral report based on one of the following questions. Work in teams and search the library or Internet for relevant pictures, facts or stories to support your points.

→ 1 Suppose you are asked to redecorate the Forbidden City, how will you accomplish the project?

→ 2 What's your understanding of the form and function of *ting*?

→ 3 What do you know about Huanghelou?

TEXT C

Chinese Religious Architecture

In the history of China, two major religions prevailed for long, long a time: Buddhism and Taoism. They played a great part in Chinese social life and influenced Chinese architecture deeply.

Chinese Buddhist Architecture

Chinese Buddhist architecture includes temples, pagodas and grottoes. Localization of Buddhist architecture started right after Buddhism was introduced into China during the Han Dynasty, interpreting Chinese architectural aesthetics and culture.

As the center of spreading Buddhism in China, temples are where Buddhist monks perform their religious life. Since many emperors believed in Buddhism, temples were built one after another, usually splendid like palaces, for many of them were built under imperial orders. In the Northern Wei Dynasty, there were more than 30,000 temples scattered in the country. Later as architectural techniques improved, glazed tiles, exquisite engravings and delicate paintings were applied in the construction of temples, which came to be more magnificent and splendid.

Chinese Buddhist architecture follows symmetric style strictly. Usually main buildings will be set on the central axis, facing the south. Annexe structures will be on the west and east flanks. Temple gate, Heavenly King Hall, the Main Hall and Sutra Library successively stands on the axis. Dorm, kitchen, dining hall, storehouse and antechamber usually cluster on the right side while left side is kept for the visitors.

Pagoda is also the main integrating part of the Buddhist architecture, with varied styles and strong local flavors. Pagoda followed Buddhism into China around the first century, and developed into pavilion-like pagoda on which one can view scenery after immediate combination with traditional Chinese architecture. Now the highest pagoda existing stands 40 meters high and enjoys a 1,400-year lifespan after survival of several earthquakes. Among the 3,000 existing pagodas, there are all-timber pagodas, brick pagodas, stone pagodas, bronze pagodas and iron pagodas.

Most Chinese pagodas are multi-storey ones. Early pagodas were usually wooden and had quadrangle, hexangle, octangle and twelve sided ichnographies. During the Sui and Tang dynasties, pagodas tended to be stone and brick. In the Liao Dynasty, solid pagodas appeared. In the Song, Liao and Jin dynasties, flower pagodas were introduced which were decorated with assorted carved flowers, honeycombed shrines, animals and Buddha and disciple sculptures, looked like flowers. Generally speaking, pagodas became more and more decorative.

Though there are various types of pagodas, they have a common structure, an underground palace. The most famous palace underground lies at the Famen Temple in Xi'an, Shaanxi.

Another Buddhist architecture is grotto complex which is caves hewn on cliff walls, usually huge projects and with exquisite engravings. It came from India with Buddhism too and boomed during the Southern and Northern Dynasties. The famous Mogao Grottoes, Yungang Grottoes and Longmen Grottoes were all carved then.

Taoist Architecture

Taoist architecture includes various structures according to different functions, categorized as palace for oblation and sacrifice, altar for praying and offering, cubby for religious service, residence for Taoist abbes and garden for visitors.

During the last period of the East Han Dynasty when Taoism was formed, Taoist ascetics mostly lived in huts and even caves in remote mountains under guidance of their philosophy of nature.

During the Jin Dynasty and the Southern and Northern Dynasties, Taoism experienced reforms and was accepted by the rulers. Many Taoist temples were set up in the capital under imperial orders. Taoist architecture reached a rather large scale then.

Taoism reached its peak during the Tang Dynasty and the Song Dynasty, when Chinese timber framed architecture, characterized by high base, broad roof and perfect integration of decoration and function, matured in all aspects. There were strict regulations on size, structure, decoration and use of color. For the 660 years, Taoism, Buddhism and Confucianism influenced each other, so that certain structures in Buddhist and Confucianist architectures were transformed into Taoist architecture. As a result, there remained similarities in designing and grouping among the three systems.

Taoist architecture applies two architectural styles—traditional style and Bagua style.

In the former style, traditional architectural layout, which is symmetric, will be applied.

Main halls will be set up on the central axis, while other religious structures on the two sides. Usually, on the northwest corner of the complex, Lucky Land to meet God will be located. Annexes like dining hall and accommodation will be located at the back or the flank of the complex.

The second is the Bagua style in which all structures surround the stove to make pills of immortality in the center according to Bagua's position request. The central axis from the south to the north is very long and structures flank the axis. The style reflects Taoist philosophy that the human cosmos follows the natural cosmos to integrate energy, *qi* and spirit.

Most Taoist architectures resort to natural topography to build towers, pavilions, lobbies and other garden structural units, decorated with murals, sculptures and steles to entertain people, fully interpreting Taoist philosophy of nature.

Taoist architectural decoration reflects Taoist pursuit of luck and fulfillment, long lifespan, and eclosion into the fairyland. Taoist architectural motifs are all meaningful. Celestial bodies mean brightness shining everywhere while landscape and rocks immortality. Folding fan, fish, narcissus, bat and deer are used to imply beneficence, wealth, celestial being, fortune and official position, while pine and cypress stand for affection, tortoise for longevity, crane for man of honor. There are many other symbols very traditional and Taoist decorations root deep in Chinese folk residential houses.

VOCABULARY

New Words

Taoism *n.* 道教，道家学说
localization *n.* 地方化
exquisite *a.* 精致的，精巧的
flavour *n.* 特色
assorted *a.* 多样的，混合的
honeycombed *a.* 蜂窝状的
Buddha *n.* 佛，佛陀
hew *v.* (hewn为过去分词) 砍，劈
oblation *n.* 供品，祭品
abbe *n.* 僧侣，道士
ascetic *n.* 修道者
Confucianism *n.* 儒教，儒家学说或思想
resort to 采取，诉诸于
eclosion *n.* 涌现
celestial *a.* 天空的，天国的
beneficence *n.* 德行，善行，仁慈
cypress *n.* 柏树

Architectural Terms

grotto *n.* 岩洞, 洞穴
annexe *n.* 附属建筑
flank *n.* 两侧, 侧翼
antechamber *n.* 前厅, 接待室, 香客室
ichnography *n.* 平面图
cubby *n.* 小房间
mural *n.* 壁画, 壁饰

Proper Names

Buddhist architecture 佛教建筑
Sutra Library 藏经室
Northern Wei Dynasty 北魏朝
Sui Dynasty 隋朝
Liao Dynasty 辽代(朝)
Song Dynasty 宋朝
Jin Dynasty 晋朝
Southern and Northern Dynasties 南北朝
Mogao Grottoes 莫高窟
Yungang Grottoes 云岗石窟
Longmen Grottoes 龙门石窟
Taoist architecture 道教建筑
Bagua style 八卦式
Lucky Land 乐土

Answer the following questions according to the text.

1. What religions influenced Chinese architecture so much?
2. What principles does Chinese Buddhist architecture follow?
3. What is the specialty in Chinese Buddhist architecture?
4. What are the various functions of Chinese Taoist architecture?
5. How many styles does Chinese Taoist architecture have?

7 Unit

Tall Buildings

Warming Up

1. Discuss with your partner about the tallest building in your city, such as the name of the building, its designer, the year of its completion, the height, etc.
2. What is the city Chicago best known for?

TEXT A

Skyscrapers

It has been stated that skyscraper and the 20th century are synonymous and there can be no doubt that the tall building is the landmark of our generation. It is a structural marvel that reaches to the heavens and embodies human goals to build ever higher. The skyscraper is the century's most stunning architectural accomplishment.

But the question of how to design the tall building still continues to taunt, disconcert, and confound practitioners. The swing in taste and style is as predictable as night and day, and we are this very moment busy rewriting the rules of skyscraper design. In the process we are not sure that the right lessons we have learned are not being discarded for the wrong ones.

A successful skyscraper solution and the art of architecture itself depend on how well the structural, utilitarian, environmental, and public roles of the tall building are resolved. Style, any style, must be intrinsic to, and expressive of, these considerations. Architecture is, above all, an expressive art.

The skyscraper has totally changed the scale, appearance, and concept of our cities and the perceptions of people in them. No doubt it will continue to do so. But it is more important today than ever that the builder and the architect consider all the factors associated with the design of a tall building and how it is incorporated into its urban setting.

Looking at the whole historical spectrum of skyscraper design, four significant phases can be identified: the functional, the eclectic, the modern, and what is currently called the postmodern, a term coined more by the media, for surely our references to modernism have not changed but have merely broadened.

It is significant that all of the most important structural solutions came early in the development of tall buildings and in a very short space of time. Because these structures were concentrated in Chicago in the two decades at the end of the last century, it was quickly acknowledged and referred to as the Chicago style.

The period from 1890 to 1920 was considered the golden age of architecture, and there have been few more masterful and original tall buildings produced than those by the architect Louis Sullivan. Running as counter current to the already emerging eclecticism, Sullivan believed that the design of the skyscraper was the translation of structure and plan into appropriate cladding and ornament and that the answers were not to be found in the rules of the past.

The eclectic phase produced some most remarkable monuments, employing many of the styles and ornamentation from the temples of Greece to the Italian Renaissance. The best examples displayed skilled academic exercises, composed with ingenuity and drama to answer the new needs and aspirations of the twentieth century. These designs so beautifully complied by architects like Raymond Hood and Cass Gilbert culminated in the famous international competition for the Chicago Tribune Tower in 1922.

▲ the Chicago Tribune Tower

This competition, which called for "the Most Beautiful and Distinctive Office Building in the World", drew more than 200 entries. The selection of the Gothic revival design by Howells and Hood prolonged the eclectic style against the concepts of the modern. For ten years modernism as pioneered by a relatively few European architects, paralleled a style that would better be termed modernistic. This style was neither pure nor revolutionary, but fused the end of the decorative eclectic style with the modernist theories and has become popularly known today as Art Deco.

The early modern or International style skyscrapers are small in number because of lack of courage on the part of the builder and a reluctance to invest in a style not yet accepted. But after the Second World War the descendants of these early modern skyscrapers, such as the McGraw-Hill Building in Manhattan, came to make up the high modern corporate style, the flat top glass boxes that have been the focal point of criticism over the past ten years.

◀ McGraw-Hill Building in Manhattan

These big buildings have taught us a hard lesson. But it is wrong that so much has been blamed on the esthetic, for such problems owe just as much to investment patterns and social upheavals. Unfortunately the minimalism of the modern esthetics let itself to the cheapest corner cutting. Since this is the most profitable route for the builder to take, it is an elegant and refined vocabulary that was quickly reduced to bottom line banality. Many are already grieving the passing; for it is structure in its purest form, enclosed in a sheer curtain of shaped and shimmering glass, that has produced some of the most innovative designs of our time.

These ideas should not be abandoned in search for ideal answers. After all, the history of the skyscraper—which is also the history of the century—is a search for identity.

VOCABULARY

New Words

synonymous *a.* 同义的
landmark *n.* 里程碑，界标
marvel *n.* 奇迹
stunning *a.* 令人吃惊的
taunt *v.* 嘲笑，讥讽
disconcert *v.* 挫败，使窘迫
confound *v.* 使困惑，使不知所措
utilitarian *a.* 实用的
intrinsic *a.* 内在的，固有的
spectrum *n.* 范围
culminate *v.* 达到顶点
fuse *v.* 熔合，混合
upheaval *n.* 动乱，剧变
minimalism *n.* 极简抽象派艺术
banality *n.* 平庸，陈腐
shimmering *a.* 闪闪发光的

Architectural Terms

cladding *n.* 涂层，敷层

Proper Names

Raymond Hood 雷蒙德·胡德
Cass Gilbert 凯斯·吉尔伯特
Chicago Tribune Tower 芝加哥论坛报大厦
Howells 豪威尔斯
Art Deco 装饰派艺术（起源于20世纪20年代的装饰和建筑艺术风格，以轮廓和色彩明朗粗犷、呈流线型和几何形为特点）
McGraw-Hill Building 麦克格劳–希尔大厦

EXERCISES

I. Match the Chinese expressions with their English equivalents.

1. 哥特复兴式	A. skyscraper
2. 装饰艺术风格	B. Italian Renaissance
3. 意大利复兴运动	C. marvel
4. 装饰，装饰品	D. the Gothic revival
5. 摩天大楼	E. eclecticism
6. 涂层，敷层	F. cladding
7. 规模	G. Art Deco
8. 奇迹	H. innovative design
9. 创新设计	I. ornamentation
10. 折中主义	J. scale

II. Decide whether the following statements are True or False.

1. A successful skyscraper solution and the art of architecture itself depend on how well the structure is resolved, disregarding utilitarian, environmental, and public roles of the tall building.
2. The skyscraper has totally changed the scale, appearance, and concept of our cities and the perceptions of people in them.
3. Sullivan believed that the design of the skyscraper was the translation of structure and plan into appropriate cladding and ornament and that the answers were to follow the rules of the past.
4. The eclecticism-dominant period produced some most remarkable monuments, employing many of the styles and ornamentation from the temples of Greece to the Italian Renaissance.
5. Because the profit is a major concern for most builders, monotony in some modern design was an innovative trend.

III. Choose the best answer to each of the following questions.

1. Since any architectural style must be intrinsic to and expressive of the considerations such as the structural, utilitarian, environmental, and public roles, architecture is, above all, an _______ art.

   A. fine   B. expressive   C. impressive   D. visual

2. Taking the whole historical spectrum of skyscraper design into consideration, four significant phases can be identified: the functional, the eclectic, the modern, and what is currently called the ______.

   A. neoclassical B. classical revival C. postmodern D. neo-modern

3. Tall structures were concentrated in ______ in the two decades at the end of the last century, and it was quickly acknowledged and referred to as the ______ .

   A. New York; Chicago style B. New York; New York style

   C. Chicago; Chicago style D. Chicago; New York style

4. Art Deco was neither pure nor revolutionary, but fused the end of the decorative ______ style with the ______ theories.

   A. eclectic; modernist B. eclectic; classical

   C. modernist; eclectic D. modernist; classical

5. The McGraw-Hill Building in Manhattan made up the high modern corporate style, the ______ top glass boxes that have been the focal point of criticism over the past ten years.

   A. domed B. pointed C. vaulted D. flat

IV. Oral task

In next class, you'll be asked to give an oral report based on one of the following questions. Work in teams and search the library or Internet for relevant pictures, facts or stories to support your points.

→ 1 What do you know about the skyscraper?

→ 2 What is your understanding of the architecture principle that the Chicago Tribune Tower follows?

→ 3 How do you understand "skyscraper… embodies human goals to build ever higher"?

TEXT B

Architectural Styles of High-Rise Building in the U.S. (I)

Historical Styles

Neoclassicism (about 1750 to 1930, with a few stray examples into the 1960s)

Neoclassicism is one of many revivals of ancient Greek and Roman styles in the history of architecture. Earlier revivals include the Romanesque, Renaissance, and Baroque styles. Neoclassical architecture began in the mid-18th century as a return to idealized and authentic classical forms, in reaction to the excesses of Baroque and Rococo interpretations of classicism.

The most common features of the style are colonnades and arches. The facades are nearly always brick or stone. The overall building design usually follows the pattern of the classical column: a pronounced base with a ceremonial entrance, a uniform shaft with little decoration, and a distinctive or pronounced top.

Neo-Gothic (about 1905 to 1930)

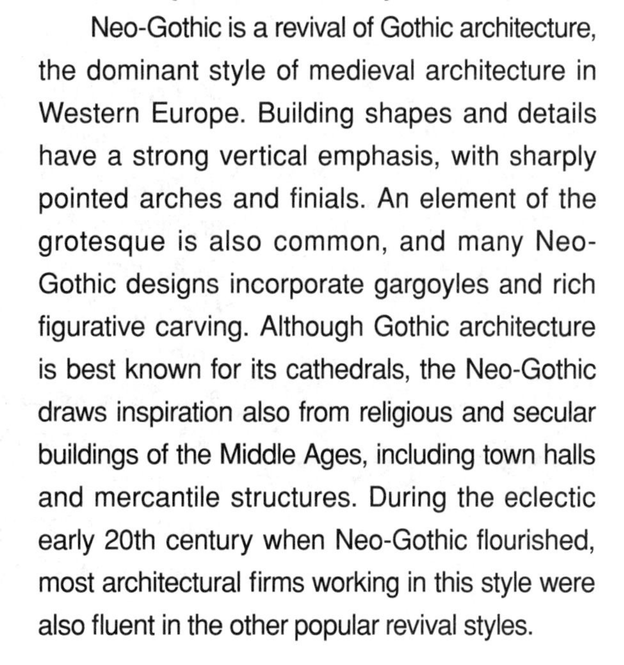

Neo-Gothic is a revival of Gothic architecture, the dominant style of medieval architecture in Western Europe. Building shapes and details have a strong vertical emphasis, with sharply pointed arches and finials. An element of the grotesque is also common, and many Neo-Gothic designs incorporate gargoyles and rich figurative carving. Although Gothic architecture is best known for its cathedrals, the Neo-Gothic draws inspiration also from religious and secular buildings of the Middle Ages, including town halls and mercantile structures. During the eclectic early 20th century when Neo-Gothic flourished, most architectural firms working in this style were also fluent in the other popular revival styles.

Renaissance Revival (about 1895 to 1930)

Renaissance revival is a branch of neoclassicism influenced by the palaces, fortresses, and public buildings of the Italian Renaissance like the Palazzo Vecchio in Florence and various Venetian landmarks.

Most buildings in this style have brick facades. Common features include towers or turrets, pyramidal roofs, castellation, large indented cornices, and rows of arched windows.

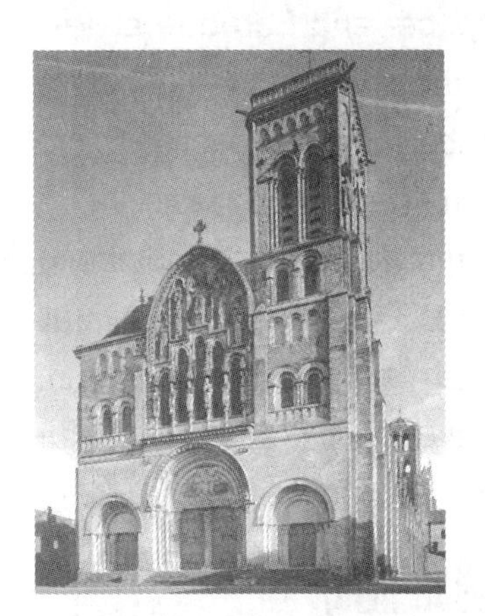

Romanesque (about 1870 to 1905)

The late 19th and early 20th century style of Romanesque is a revival of an early medieval style, which was in turn a revival of Roman architecture. This was one of the most popular forms of architecture in the United States during the 1880s, and along with the Chicago school it was the first style applied to tall buildings. Many courthouses and public buildings were built in Romanesque, even in small rural towns.

Distinguishing features include turrets, rounded arches, hipped or pointed roofs, and very heavy rusticated stonework. Proportions in this style tend to run large, both in the overall building form and in the size of the details.

Transitional Styles

Art Deco/Art Modern (about 1920 to 1940)

Art Deco and Art Modern are two ends of a continuum, forming a style with roots in the verticality of Gothic architecture but leaning toward the simplicity of modernism. As its name implies, Art Deco uses more decoration while Art Modern buildings tend to have smoother, streamlined shapes. Both Deco and Modern use setbacks to reduce building mass and to emphasize verticality. Unlike "Wedding Cake" buildings, their shapes recede from the street gracefully, not in tiers but in gentler and more carefully positioned steps. Limestone is the most common cladding material, with brick facades common in Art Deco.

Chicago School (about 1885 to 1915)

The "Chicago School of Architecture" was a proto-modernist style which arose during the building boom after the Chicago Fire. The style is a major step in the direction of simplified modern architecture, and although it incorporates many features of historical styles the ornament is subordinated to the overall structural scheme. The style encompasses the first skyscrapers, and in many buildings the facade depicts nothing more than the rectangular steel grid underneath. Buildings in this style were built in various cities, mostly in the Midwest but even in New York. Its influence was very strong in industrial architecture, and many early factories and warehouses fall into this category of design.

(To be continued)

VOCABULARY

New Words

stray *a.* 孤立的，零星的
mercantile *a.* 贸易的，经商的
fortress *n.* 堡垒，要塞
indented *a.* 锯齿状的
courthouse *n.* 县政府大楼
transitional *a.* 过渡时期的
rusticate *v.* 使成粗面石工
continuum *n.* 连续体
streamlined *a.* 流线型的，现代型的
proto-modernist *a.* 典型现代主义的
depict *v.* 描绘；描画

Architectural Terms

grotesque *n.* （哥特式建筑的）怪诞饰
gargoyle *n.* （哥特式建筑的）怪兽状的滴水嘴
town hall 市政厅
turret *n.* 角楼，塔楼
castellation *n.* 城堡形建筑
cornice *n.* 檐口，飞檐
hipped *a.* 有斜脊的
setback *n.* （墙壁上部厚度减低形成的）壁阶，缩进
grid *n.* 棋盘式街道布局

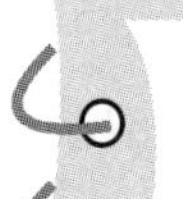

Proper Names

Palazzo Vecchio 韦基奥宫（意大利佛罗伦萨）
the Chicago Fire 芝加哥大火

EXERCISES

I. Translate the following English expressions into Chinese.

1. classicism ____________
2. colonnade ____________
3. arch ____________
4. facade ____________
5. town hall ____________
6. cornice ____________
7. turret ____________
8. medieval architecture ____________
9. grid ____________
10. pyramidal roof ____________

II. Match each statement with the correct architectural style. There are three extra statements which you do not need to use.

a. Romanesque ☐   b. Neoclassicism ☐   c. Chicago school ☐

d. Renaissance revival ☐   e. Art Deco ☐   f. Neo-Gothic ☐

1. The facade depicts nothing more than the rectangular steel grid underneath.
2. Building shapes and details have a strong vertical emphasis, with sharply pointed arches and finials.
3. Their shapes recede from the street gracefully, using setbacks to reduce building mass and to emphasize verticality.
4. Common features include limestone or brick facades, stone framing around entrances, smooth stone columns, and rounded building or window edges.
5. Common features include towers or turrets, pyramidal roofs, castellation, large indented cornices, and rows of arched windows.
6. Designs which embrace the roughness of concrete or the heavy simplicity of its natural forms.
7. Colonnades and arches are the most common features of the style.
8. Common features include columns, pyramids, arches, obelisks, unusual or attention-getting shapes and rooflines, and combinations of stone and glass on the facade.
9. Distinguishing features include turrets, rounded arches, hipped or pointed roofs, and very heavy rusticated stonework.

III. Oral task

In next class, you'll be asked to give an oral report based on one of the following questions. Work in teams and search the library or Internet for relevant pictures, facts or stories to support your points.

→ 1 What do you know about the neoclassical style and the architecture of this style?

→ 2 What is the defined feature of Gothic architecture?

→ 3 What do you know about Art Deco and Art Modern?

TEXT C

Architectural Styles of High-Rise Building in the U.S. (II)

Modern Styles

Brutalism (about 1950 to 1985)

Although the word Brutalism comes from the French word for rough concrete, a sense of brutality is also suggested by this style. Brutalist structures are heavy and unrefined with coarsely molded surfaces, usually exposed concrete. Their highly sculptural shapes tend to be crude and blocky, often colliding with one another.

The line between brutalism and ordinary modernism is not always clear since concrete buildings are so common and run the entire spectrum of modern styles. Designs which embrace the roughness of concrete or the heavy simplicity of its natural forms are considered brutalist. Other materials including brick and glass can be used in brutalism if they contribute to a block-like effect similar with the strongly articulated concrete forms of early brutalism.

Early Modernism (about 1930 to 1955)

This style represents the early stages of the modern style, in which buildings often retain traces of the Art Modern or neoclassical styles. Buildings in this genre frequently have streamlined shapes or facade designs, and are very often broken into distinct masses or wings with courtyards.

Common features include limestone or brick facades, stone framing around entrances, smooth stone columns, and rounded building or window edges. Windows are frequently arranged in strong horizontal bands, or are punched into the facade at regular intervals.

Futurism (about 1955 to the present)

Futurism is a broad trend in modern design which aspires to create architecture of an imagined future, normally thought to be at least 10 years into the future. The beginnings of futurism go back to the visionary drawings of Italian architect Antonio Sant'Elia. Early features of Futurism included fins and ledges, bubble shapes and sweeping curves. The style has been reinterpreted by different generations of architects across several decades, but is usually marked by striking shapes, clean lines, and advanced materials.

International Style (about 1930 to 1980, with examples continuing into the present)

The International style is the purest and most minimal form of modernism. It originated in a number of movements from Germany and the Netherlands in the 1920s, especially the Bauhaus. Its designs are generally simple prismatic shapes, with flat roofs and uniform arrangements of windows in bands or grids.

The most common materials in International style buildings are glass, steel, aluminum, concrete, and sometimes brick infill. Plaster, travertine marble and polished stone are common on the interiors.

The leader of the Bauhaus school and a founder of the International style was the architect Walter Gropius. Another Bauhaus architect, Ludwig Mies van der Rohe, is the most famous and influential figure in the movement.

Modernism (about 1940 to the present)

The most common building style worldwide, standard modernism has evolved from utilitarian forms introduced in the 19th century. Modernist buildings are generally simple in design and lack any applied ornament. Their architecture is basically a modification of the International style but is less strict in its geometry.

After about 1960 modernism began to play more freely with shapes and structures, producing a wider variety of designs including cylindrical buildings, sloping roofs, and unusual shapes. This trend runs parallel to postmodernism, which rebelled against the strictness of modernism by reviving historical tropes; but during this period the aesthetic and economic advantages of simplicity kept modernism alive in all parts of the world.

Postmodernism (about 1970 to the present)

Postmodern architecture is a counteraction to the strict and almost universal modernism of the mid-20th century. It reintroduces elements from historical building styles, although usually without their high level of detail. Common features include columns, pyramids, arches, obelisks, unusual or attention-getting shapes and rooflines, and combinations of stone and glass on the facade.

Postmodernism ranges from conservative imitations of classical architecture to flamboyant and playfully outrageous designs. As the style became mainstream, many buildings with a modern form assimilated postmodern devices into small parts of their designs.

Structural Expressionism (about 1975 to the present)

Also called "high-tech modernism", structural expressionism is a specific branch of advanced modernism in which buildings display their structural elements visibly inside and out. The larger design features are liberated by the possibilities of engineering, while detailing is generally faithful to the principles of the International style. Common features include detached frames, exposed truss work, and highly complex shapes requiring unusual engineering.

Structures in this style tend to be metallic, in contrast to the older brutalist style which usually employs concrete. Precedents of structural expressionism include modern buildings like the John Hancock Center and U.S. Steel Tower.

VOCABULARY

New Words

unrefined *a.* 不精细的，粗糙的
coarsely *adv.* 粗糙地
crude *a.* 不精细的
collide *v.* 碰撞，冲突
articulated *a.* 具有表现力的
aluminum *n.* 铝
infill *n.* 填充
travertine *n.* 石灰华
ornament *n.* 装饰品
slope *v.* 倾斜
trope *n.* 比喻
counteraction *n.* 消除，抵消
flamboyant *a.* 艳丽的
outrageous *a.* 粗暴的
assimilate *v.* 同化，吸收

Architectural Terms

brutalism 粗野主义(一种展示未经装饰的巨大构件以表示结实和粗犷力量的建筑风格)
futurism 未来主义
prismatic *a.* 棱柱型的
cylindrical *a.* 圆柱体的
roofline *n.* 屋顶轮廓
structural expressionism 结构表现主义
truss work (支撑屋顶，桥等的)构架

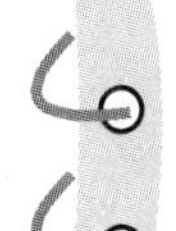

Proper Names

Antonio Sant'Elia 安东尼奥·桑塔埃利亚(意大利著名建筑设计师)
John Hancock Center 约翰·汉考克大厦(美国芝加哥)
U.S. Steel Tower 美国钢铁大厦(芝加哥)

Answer the following questions according to the text.

1. What designs are considered brutalist?
2. Give a brief description of the common features of the style of early modernism architecture.
3. How many years is the style of futurism architecture normally thought to be into the future?
4. When did modernism begin to play more freely with shapes and structures?
5. What is structural expressionism also called?

Unit 8

Architectural Wonders

Warming Up

1. How do you view the relationship between architecture and civilization?
2. Can you name some ancient wonders of architecture?

TEXT A

Seven Wonders of the Ancient World

In ancient times, Seven Wonders of the World were works of art and architecture regarded by ancient Greeks and Romans as the most extraordinary structures of antiquity.

The Great Pyramids of Giza, Egypt

Built in the 3rd millennium BC, the Great Pyramids are located at Giza, Egypt near Cairo City. Among them Khufu's Great Pyramid is the most remarkable. It is 756 feet long on each side, 450 high and is composed of 2.3 million blocks of stone, each averaging 2.5 tons in weight. Despite the makers' limited surveying tools no side is more than 8 inches different in length than another, and the whole structure is perfectly oriented to the points of the compass. Until the 19th century it was the tallest building in the world and, at the age of 4,500 years, it is the only one of the famous Seven Wonders of the Ancient World that still stands.

The Hanging Gardens of Babylon

Built by King Nebuchadnezzar II about 600 BC, the Hanging Gardens of Babylon was a mountain-like series of palace with planted terraces. The garden was within the ancient city of Babylon, which must have been a wonder to the travelers' eyes. "In addition to its size," wrote Herodotus, a historian in 450 BC, "Babylon surpasses in splendor any city in the known world." Herodotus claimed the outer walls were 56 miles in length, 80 feet thick and 320 feet high, wide enough to allow a four-horse chariot to turn. The inner walls were "not so thick as the first, but hardly less strong". Inside the walls were fortresses and temples containing immense statues of solid gold. Rising above the city was the famous Tower of Babel, a temple to the god Marduk, which seemed to reach to the heavens.

The Statue of Zeus at Olympia, Greece

The seated statue was built to be in honor of Zeus, the king of the Greek's gods in the 5th century BC. It was created by Greek sculptor Phidias. It was seated inside the Temple of Zeus at Olympia, where the original Olympic Games were initiated. The Zeus figure's skin was composed of ivory and the beard, hair and robe of gold. Construction was by the use of gold and ivory plates attached to a wooden frame. Because the weather in Olympia was so damp, the statue required care so that the humidity would not crack the ivory. For this purpose it was constantly treated with oil kept in a special pool in the floor of the temple. It is said that for centuries the descendants of Phidias held the responsibility for this maintenance of the statue.

The Temple of Artemis at Ephesus, Asia Minor

At about the 4th century BC an architect named Chersiphron was engaged to build a large temple for Artemis, the goddess of fertility, favored by the citizens of Ephesus, which had become a major trade port of Asia Minor by 600 BC. The building was designed with high stone columns. This temple didn't last long and was destroyed in 550 BC during the fighting of conquering Ephesus and the other Greek cities of Asia Minor by King Croesus of Lydia. The architect who rebuilt it was thought to be Theodorus. The rebuilt temple was 300 feet in length and 150 feet wide with an area four times the size of the former. More than one hundred stone columns supported a massive roof. The new temple was the pride of Ephesus until 356 BC when a tragedy struck—the temple was burnt down by fire.

The Mausoleum of Halicarnassus, Asia Minor

The tomb was built for Mausolus, the governor (ruler) of Caria in the 4th century BC and the term "mausoleum" came from the name of its owner. Made mostly of marble, the 140-foot high structure can be divided into three sections equally, the lowest rose as a square, tapering block and was covered with relief sculpture showing action scenes from Greek myth: the battle of the Centaurs with the Lapiths and the Greeks in combat with the Amazons, a race of warrior women. Above this section thirty-six slim columns, nine per side rose for another third of the height. Standing in between each column was another statue. Behind the columns was a solid block that carried the weight of the tomb's massive roof. The roof, which comprised most of the final third of the height, was in the form of a stepped pyramid. On top was the tomb's notable work of sculpture: four massive horses pulling a chariot in which Mausolus and his wife Artemisia drove. The Mausoleum overlooked the city of Halicarnassus for many centuries and it stood above the city ruins for some 17 centuries.

The Colossus of Rhodes

The original bronze colossus of the Greek sun god Helios stood over 2,000 years ago (280 BC) at the entrance to a busy harbor on the Island of Rhodes. Like the Statue of Liberty, this colossus was also built as a celebration of freedom. This amazing statue, standing the same height from toe to head as the modern colossus, was one 110 feet high and stood upon a 50 feet pedestal near the harbor mole. Although the statue has been popularly depicted with its legs spanning the harbor entrance so that ships could pass beneath, it was actually posed in a more traditional Greek manner: nude, wearing a spiked crown, shading its eyes from the rising sun with its right hand, while holding a cloak over its left. Located off the southwestern tip of Asia Minor where the Aegean Sea meets the Mediterranean, the island of Rhodes was an important economic center in the ancient world.

The Pharos of Alexandria

The 3th century BC lighthouse was built on the island of Pharos near Port Alexandria, Egypt and soon the building itself acquired the name "Pharos".

The 20th-century skyscraper-like structure consisted of three stages, each built on top of the lower. The building was constructed of marble blocks with lead mortar. The lowest level was probably more than 200 feet in height and 100 feet square, shaped like a massive box. The upper portion of the building contained hundreds of storage rooms. Inside this section was a large spiral ramp that allowed materials to be pulled to the top in horse-drawn carts. Staircases allowed visitors and the keepers to climb to the beacon chamber. Above this section was an eight-sided tower. On top of the tower was a cylinder that extended up to an open cupola where the fire that provided the light burned. There, according to reports, a large curved mirror, perhaps made of polished metal, was used to project the fire's light into a beam. It was said ships could detect the light from the tower at night or the smoke from the fire during the day up to 100 miles away. On the roof of the cupola was a large statue of Poseidon.

The list of man-made wonders was first compiled by a Hellenistic traveler in the 2nd century BC.

VOCABULARY

New Words

splendor *n.* 壮丽，壮观，辉煌
ivory *n.* 象牙
spiked crown 带尖芒的王冠
cloak *n.* 斗篷，外衣
pharos *n.* 灯塔
beacon chamber (灯塔) 照明室，信号室
beam *n.* 光束
Hellenistic *a.* 希腊风格的，希腊文化的

Architectural Terms

tapering *a.* 锥形、锥体的
relief *n.* 浮雕
colossus *n.* 巨像
pedestal *n.* 基座，底座
span *v.* 横跨
lead mortar 铅灰泥
ramp *n.* 斜坡，坡道
cupola *n.* 小穹顶

Proper Names

Nebuchadnezzar II 尼布甲尼撒二世（古巴比伦国王，攻占耶路撒冷，建空中花园）
Herodotus 希罗多德（古希腊著名历史学家）
Tower of Babel 巴别之塔（圣经中所传的巴比伦通天塔）
Marduk 马杜克（古代巴比伦人的主神，原为巴比伦的太阳神）
Olympia 奥林匹亚（希腊南部平原，用以祭拜宙斯的宗教中心，古代奥林匹克运动会的遗址）
Phidias 菲迪亚斯（古希腊雅典雕塑家，奥林匹亚的宙斯雕像是其杰作）
Artemis 阿耳忒弥斯女神（希腊神话中狩猎女神和月神）
Ephesus 以弗所（小亚细亚西岸的贸易城市，以阿耳忒弥斯神庙而闻名）
Chersiphron 切西弗龙（古希腊著名建筑师，建造了阿耳忒弥斯神庙）
Croesus 克利萨斯（吕底亚王国的末代国王，以富有著称）
Lydia 吕底亚（小亚细亚中西部一古国，以其富有和奢华而闻名）
Theodorus 西奥多罗斯（古希腊建筑师，重建了阿耳忒弥斯神庙）
Halicarnassus 哈利卡那苏斯（位于小亚细亚今天土耳其境内西南部爱琴海海滨的希腊文化古城）
Mausolus 摩索拉斯王（波斯帝国卡里亚省的总督，其陵墓非常壮观）
Caria 卡里亚（小亚细亚西南部濒临爱琴海的一古老地区，曾为波斯帝国行省）
Centaur （希腊神话）人首马身怪
Lapith 拉毗士（古希腊塞萨利地区的部落）
Amazon 亚马逊族（相传曾居住在黑海边强壮的女战士一族）
Rhodes 罗得岛（位于爱琴海畔东南部）
Helios（希腊神话）太阳神赫利俄斯（相当于罗马神话的阿波罗）
Statue of Liberty 自由女神像（纽约）
Aegean Sea 爱琴海
Mediterranean 地中海
Pharos 法罗斯岛（位于埃及亚历山大港的地中海上，后来指代灯塔）
Alexandria 亚历山大港（埃及北部港市）
Poseidon（希腊神话）海神波塞冬

EXERCISES

I. Match the English expressions with their Chinese equivalents.

1. spiked crown	A. 巨像
2. relief	B. 灯塔
3. colossus	C. 横跨
4. pedestal	D. 小穹顶
5. span	E. 斜坡，坡道
6. pharos	F. 带尖芒的王冠
7. lead mortar	G. 铅灰泥
8. ramp	H. 圆筒，圆柱体
9. cylinder	I. 浮雕
10. cupola	J. 基座，底座

II. Decide whether the following statements are True or False.

1. Until the 20th century the Pyramid of Khufu at Giza was the tallest building in the world and, at the age of 4,500 years, it is the only one of the famous Seven Wonders of the Ancient World that still stands.
2. The seated statue inside the Temple of Zeus at Olympia was created by Greek sculptor Phidias.
3. The Temple of Artemis was designed with high stone columns firstly by Theodorus, a great Greek architect.
4. The term "mausoleum" comes from the name of Mausolus, a governor (ruler) of Caria, an ancient region of Asia Minor.
5. At the top part of the pharos, a large plain mirror, perhaps made of polished metal, was used to project the fire's light into a beam.

III. Choose the best answer to each of the following questions.

1. The Zeus figure's skin was composed of ______ and the beard, hair and robe of ______.
   A. marble; gold B. marble; silver C. ivory; gold D. ivory; silver
2. At about the 4th century BC an architect named Chersiphron was engaged to build a large temple for Artemis, the goddess of ______.
   A. wisdom B. beauty C. power D. fertility

3. The lowest part of Mausolus' tomb was featured as covered with ________ showing action scenes from Greek myth.
   A. paintings B. handwritings C. relief sculpture D. sculpture
4. Like the Statue of Liberty, the colossus of Helios was also built as a ________.
   A. celebration of freedom B. warfare
   C. commercial sign D. worship of god
5. The 20th-century skyscraper-like Pharos at Alexanderia consisted of ________ stages, each built on top of the lower.
   A. two B. three C. four D. multiple
6. The list of the seven man-made wonders was first recorded by a(n) _______ traveler in the 2nd century BC.
   A. Ancient Egyptian B. Arabian C. Babylonian D. Ancient Greek

IV. Writing task

Search the library or Internet for information about Tilt (the tower of Pisa) and write an essay about it. The essay should include:

→ 1 the brief history of Tilt;

→ 2 the brief introduction of Tilt (its location, height, its architectural style, etc.)

→ 3 its significance in architecture.

TEXT B

New Seven Wonders of the World

Inspired by the Seven Wonders of the Ancient World, the World Heritage Organization decides to crown seven new world wonders among the candidates. July 7th, 2007 saw the nomination of the new seven world wonders in Lisbon, Portugal. Gloriously crowned, the following were chosen as the New Seven Wonders of the World from among the 20 candidates (excluding the Great Pyramids of Giza) by the people all over the world.

The Great Wall of China

Snaking along the mountains in the southern part of the Mongolian Plain, the Great Wall (or Walls) was built over centuries, beginning as early as 500 BC. During the Qin Dynasty, many walls were joined and re-enforced for greater strength. In places, the massive walls are as tall as 29.5 feet (9 meters). The Great Wall of China was built to link existing fortifications into a united defense system. It is the largest man-made monument ever to have been built and it is disputed that it is the only one visible from space. No one is sure exactly how long the Great Wall of China is. Many say that the Great Wall extends some 3,700 miles (6,000 kilometers).

Petra, Jordan

On the edge of the Jordan Desert, the strikingly beautiful city of Petra was the capital of the Nabataean Empire (9 BC to AD 40). Masters of water technology, the Nabataeans provided their city with great tunnel constructions and water chambers. A theater, modeled on Greek-Roman prototypes, had space for an audience of 4,000. Today, the Palace Tombs of Petra, with the 42-meter-high Hellenistic temple facade, are impressive examples of Middle Eastern culture. The rose-red city was lost to the Western World from about the 14th century until it was discovered in 1812.

The Colosseum in Rome, Italy

The Roman emperors Vespasian and Titus built the Colosseum in central Rome between AD 70 and 82. At least 50,000 spectators—and possibly many more—could sit in the enormous 3-storey building. This great amphitheater in the center of Rome was built to give favors to successful legionnaires and to celebrate the glory of the Roman Empire. Its design concept still stands to this very day, and virtually every modern sports stadium some 2,000 years later still bears the irresistible impression of the Colosseum's original design. Today, through films and history books, we are even more aware of the cruel fights and games that took place in this arena, all for the joy of the spectators.

Machu Picchu in Peru

In the 15th century, the Inca constructed the small city in the clouds on the mountain known as Machu Picchu ("Old Mountain"). Beautiful and remote, the buildings were constructed of finely cut white granite blocks and no mortar was used. This extraordinary settlement lies halfway up the Andes Plateau, deep in the Amazon jungle and above the Urubamba River. It was probably abandoned by the Incas because of a smallpox outbreak and, after the Spanish defeated the Incan Empire, this legendary city of the Inca was almost lost to explorers until the early 1900s.

Chichén Itzá in Yucatan, Mexico

Located about 90 miles from the coast in the northern Yucatan peninsula, Chichén Itzá, the most famous Mayan temple, served as the political and economic center of the Mayan civilization. Its various structures—the pyramid of Kukulkan, the Hall of the Thousand Pillars, and the Playing Field of the Prisoners—can still be seen today and are demonstrative of an extraordinary commitment to architectural space and composition. The pyramid itself was the last, and arguably the greatest, of all Mayan temples.

The Taj Mahal in Agra, India

Made entirely of marble and standing in formally laid-out walled gardens, the glistening white Taj Mahal is regarded as the most perfect jewel of Muslim art in India. This immense mausoleum was built on the orders of Shah Jahan, the fifth Muslim Mogul emperor, to honor the memory of his beloved late wife. Beautifully symmetrical, each element of the Taj Mahal is independent, yet perfectly integrated with the structure as a whole. The master architect was Ustad 'Isa. Some 20,000 workers spent 22 years on the construction. The emperor was consequently jailed and, it is said, could then only see the Taj Mahal out of his small cell window.

Christ Redeemer Statue in Rio de Janeiro, Brazil

The Art Deco-style Christ Redeemer statue towers atop the Corcovado Mountain overlooking Rio de Janeiro. Designed by Brazilian Heitor da Silva Costa and created by French sculptor Paul Landowski, it is one of the world's best-known monuments. The statue is 125 feet (38 meters) tall, including the pedestal. The pedestal contains a chapel large enough for 150 worshippers. The statue took five years to construct and was inaugurated on October 12, 1931. It has become a symbol of the city and of the warmth of the Brazilian people, who receive visitors with open arms.

VOCABULARY

New Words

re-enforced *a.* 加固的
legionnaire *n.* 军团士兵
demonstrative *a.* 用于表明或说明的，显示出的
glistening *a.* 闪耀的，光辉的
jail *v.* 关押，监禁
redeemer *n.* 救赎者，救世主
worshipper *n.* 礼拜者，崇拜者

Architectural Terms

arena *n.* 竞技场（古罗马圆形露天竞技场的中心区域）
laid-out *a.* 设计的，布局的

Proper Names

World Heritage Organization 世界遗产组织（联合国教科文组织下属机构）
Lisbon 里斯本（葡萄牙首都和最大的城市）
Mongolian Plain 蒙古草原
Petra 佩特拉古城（约旦）
Nabataean Empire 纳比第安帝国
Vespasian 韦斯帕西恩（古罗马皇帝，开始营建古罗马圆形大竞技场）
Titus 提图斯（古罗马皇帝，统治期间继续修建罗马圆形大竞技场）
Machu Picchu 马丘比丘古城（遗迹，坐落在秘鲁境内安第斯高原上）
Andes Plateau 安第斯高原（南美洲）
Urubamba River 乌鲁班巴河（秘鲁境内的河，发源于安第斯山脉）
Yucatan 尤卡坦半岛（中美洲北部）
Ustad 'Isa 乌斯塔德·伊萨（泰姬陵的主要建筑设计师）
Corcovado Mountain 科尔瓦多山（毗邻里约热内卢市）
Rio de Janeiro 里约热内卢市（巴西东南部港市）
Heitor da Silva Costa 黑托·达·谢尔瓦·科斯塔（巴西设计师）
Paul Landowski 保罗·朗多斯基（法国雕塑家）

EXERCISES

I. Translate the following English into Chinese and Chinese into English.

1. laid-out ______________
2. prototype ______________
3. the Colosseum ______________
4. amphitheater ______________
5. arena ______________
6. 建筑物的正面 ______________
7. 礼拜者 ______________
8. 灰泥 ______________
9. 装饰艺术风格 ______________
10. 要塞，防御工事 ______________

II. Decide whether the following statements are True or False.

1. The Great Wall of China was built to link existing fortifications into a united defense system.
2. Virtually every modern sports stadium some 2,000 years later still bears the irresistible impression of the original design of the Colosseum in Rome.
3. Chichén Itzá, the most famous Incan temple, served as the political and economic center of the Incan civilization.
4. The immense mausoleum Taj Mahal was built on the orders of Shah Jahan, the fifth Muslim Indian emperor, to honor the memory of his beloved late wife.
5. Christ Redeemer Statue has become a symbol of the city Rio de Janeiro and of the warmth of the south American people, who receive visitors with open arms.

III. Choose the best answer to each of the following questions.

1. Inspired by the Seven Wonders of the Ancient World, the _________ decides to crown seven new world wonders among the candidates.
   A. World Trade Organization B. Asian Economic Cooperation Organization
   C. World Heritage Organization D. North Atlantic Treaty Organization
2. The strikingly beautiful city of Petra was the capital of the ________ Empire.
   A. Nabataean B. Persian C. Roman D. Macedonian
3. The minimum capacity of _________ spectators—and possibly many more—could sit in the enormous ________ -storey Colosseum at Rome.
   A. 50,000; four B. 50,000, three C. 30,000, four D. 30,000, three

4. Machu Picchu, the extraordinary settlement lies halfway up the _______, deep in the Amazon jungle and above the Urubamba River.

   A. Rocky Plateau　　B. Mongolian Plateau
   C. Andes Plateau　　D. Tinetan Plateau

5. Made entirely of _________ and standing in formally laid-out walled gardens, the glistening white Taj Mahal is regarded as the most perfect jewel of Muslim art in India.

   A. granite　　B. concrete　　C. stone　　D. marble

III. Oral task

In next class, you'll be asked to give an oral report based on one of the following questions. Work in teams and search the library or Internet for relevant pictures, facts or stories to support your points.

→ 1 Which wonder of the seven do you prefer and why?

→ 2 Why did these wonders win out from among so many candidates?

→ 3 How can the New Seven Wonders of the World be protected effectively?

TEXT C

The 13 Candidates of New World Wonders

The below are the 13 finalists in the campaign for the New Seven Wonders of the World but fail to be on the list.

Timbuktu in Mali, West Africa

Founded by nomads, the city of Timbuktu became legendary for its wealth. The name Timbuktu has taken on mythic meaning, suggesting a place that is very far away. The real Timbuktu lies in Mali, in West Africa. During medieval times, Timbuktu became a center for wealth, culture, art, and higher learning. That splendor is reflected today in Timbuktu's fascinating Islamic architecture.

The Acropolis in Athens, Greece

Much of the original Acropolis, including the Older Parthenon, was destroyed in 480 BC when Persians invaded Athens. Many temples were rebuilt during the Golden Age of Athens (460-430 BC) when Pericles was the ruler. Phidias, a great Athenian sculptor, and two famous architects, Ictinus and Callicrates, played key roles in the reconstruction of the Acropolis. Today, the Parthenon is an international symbol of Greek civilization and the temples of the Acropolis have become some of the world's most famous architectural landmarks.

The Statue of Liberty in New York, USA

French sculptor Frederic Auguste Bartholdi designed the Statue of Liberty, which was a gift from France to the United States for its centennial of independence. The Statue of Liberty was completed on October 28, 1886 and it is situated on the tip of Long Island, New York Harbor.

Stonehenge in Salisbury, United Kingdom

Stonehenge is made from 150 huge

rocks set in a circular pattern on the Salisbury Plain in southern England. It is considered to be a prehistoric religious center and also an astronomical device. Most of Stonehenge was built in about 2000 BC.

Sydney Opera House, Australia

Jørn Utzon began work on the Sydney Opera House in 1957, but the modern expressionist building stirred great controversy. The Sydney Opera House wasn't completed until 1973, under the direction of Peter Hall.

The Kremlin and St. Basil's Cathedral in Moscow, Russia

The Kremlin in Moscow is the symbolic and governmental center of Russia. Cathedral Square in the Kremlin has some of Russia's most important architecture.

Just outside the Kremlin Gates is St. Basil's Cathedral which is a carnival of painted onion domes in the most expressive of Russo-Byzantine traditions. St. Basil's was built between 1554 and 1560 and reflects the renewed interest in traditional Russian styles during the reign of Ivan IV (the Terrible) who built it to honor Russia's victory over the Tatars at Kazan.

Neuschwanstein Castle in Schwangau, Germany

With towering white turrets, Neuschwanstein Castle is not a medieval fortress but a fanciful 19th-century palace built by and for Ludwig II of Bavaria. Ludwig II died before the castle was fully completed. Eventually the graceful building was inspired by Wagner opera and his romanticism and was named Neuschwanstein, which means the New Castle of Swan.

Eiffel Tower in Paris (La Tour Eiffel)

Built in 1889 originally for the 1889 World Fair, in commemoration of the 100th anniversary of the French Revolution, Eiffel Tower in Paris is perhaps the most famous example of the new use for metal since before. The Tower is named after Gustave Eiffel, the French architect, designer, and engineer. The Eiffel Tower is the tallest building in Paris, and reigned for 40 years as the tallest in the world. The metal lattice-work, formed with very pure structural iron, makes the building both extremely light and able to withstand tremendous wind forces.

Hagia Sophia in Istanbul, Turkey (Ayasofya)

The English name for Hagia Sophia is Divine Wisdom. In Latin, the cathedral is called Sancta Sophia. But by any name, Hagia Sophia is an architectural treasure with remarkable examples of Byzantine mosaics. Christian and Islamic art combine in the Hagia Sophia. Hagia Sophia was a great Christian cathedral until the mid-1400s. After the Turk's conquest of Constantinople, the Hagia Sophia became a mosque. Then, in 1935, the Hagia Sophia became a museum.

Kiyomizu Temple in Kyoto, Japan

The word Kiyomizu can refer to several Buddhist temples, but the most famous is the Kiyomizu Temple in Kyoto. In Japanese, *kiyoi mizu* means pure water. Kyoto's Kiyomizu Temple was constructed in 1633 on the foundations of a much earlier temple. A waterfall from adjacent hills tumbles into the temple complex. Leading into the temple is a wide veranda with hundreds of pillars.

Easter Island Statues in Chile

Easter Island is an isolated island owned by Chili and located about 2,000 miles (3,200

km) from Chile. Polynesians traveled to Easter Island and formed a community that flourished between AD 1,000 and 1,500. During this time, they carved more than 800 statues, or Moai, from porous volcanic rock. The Moai of Easter Island stand as tall as 33 feet (10 meters) and weigh many tons. Some faces were decorated with coral eyes. Archaeologists speculate that the Moai represented a god, a mythical creature, or a revered ancestor.

Angkor Wat, Cambodia

Built by Suryavarman II between 1113 and 1150, Angkor Wat is the largest sacred temple and palace complex in Angkor city, the capital city of the Khmer civilization in Cambodia. Beautiful and ornate temples extend 40 miles (64 km) around the Cambodian village of Siem Reap. Ranging from simple brick towers to complicated stone structures, the temple complex is arguably the largest sepulcher in the world. The site is known best for its stunning sculpted murals on the interior walls of the palace and mortuary.

Alhambra Palace in Granada, Spain

Perched on a hilly terrace on the southern edge of Granada, Spain, Alhambra is an ancient palace and fortress complex with stunning frescoes and interior details. The Alhambra Palace was first constructed in the mid-1300s by the Moors and later renovated and modified in the 16th century. As a result, European features mingle with some of the finest examples of Moorish architecture in the world.

VOCABULARY

New Words

finalist *n.* 决赛选手或事物
nomad *n.* 游牧民
centennial *n.* 一百周年纪念
carnival *n.* 狂欢，欢聚
conquest *n.* 征服
tumble *v.* 翻滚，流下
porous *a.* 多孔的，渗水的
volcanic *a.* 火山的
coral *n.* 珊瑚
sepulcher *n.* 埋葬所，宗教圣物储藏所
mortuary *n.* 停尸房，太平间
perch *v.* 栖息，坐落
renovate *v.* 革新，刷新

Architectural Terms

expressionist *a.* 表现主义的
onion dome 洋葱头形屋顶（多见于俄罗斯东正教教堂）
Russo-Byzantine 俄式拜占庭风格的
lattice-work *n.* 格构，格构制品

Proper Names

Timbuktu 廷巴克图（马里中部城市）
Mali 马里（西非国家）
Pericles 伯里克利（古雅典领袖，因其推进了雅典民主制并下令建造帕台侬神庙而著名）
Ictinus 伊克蒂诺（希腊建筑师，雅典卫城的主要设计师）
Callicrates 卡立克拉特（希腊建筑师，雅典卫城的主要设计师）
Frederic Auguste Bartholdi 弗雷德里克·奥古斯特·巴托尔蒂（法国著名雕刻家，自由女神像的创造者）
Stonehenge 巨石阵（英国南部索尔兹伯里附近的一组立着的石群）
Salisbury Plain 索尔兹伯里平原
Jørn Utzon 约翰·伍钟（20世纪丹麦最重要的建筑设计师，其最著名的设计是悉尼歌剧院）
Kremlin 克里姆林宫（历代俄罗斯沙皇的宫邸，现为俄国政府办公所在地）
St. Basil's Cathedral 圣巴兹尔大教堂（坐落在莫斯科红场，为俄拜占庭式建筑著名代表作）
Cathedral Square 教堂广场（在克里姆林宫内）
Ivan IV 伊凡四世（伊凡雷帝，俄罗斯沙皇，带领俄罗斯人摆脱鞑靼人的统治，以暴烈著称）
Tatar 鞑靼人
Kazan 喀山（俄罗斯欧洲部分的东部城市，位于莫斯科东部，乌拉尔河上）
Neuschwanstein Castle 新天鹅城堡（坐落在德国境内巴伐利亚省的施旺高）
Schwangau 施旺高（德国巴伐利亚省南部

一地区，临近图森镇）
Ludwig II 路德维希二世（德国巴伐利亚的末代君主，修建了新天鹅城堡）
Bavaria 巴伐利亚（德国南部的一大省）
Wagner 瓦格纳（全名理查德·瓦格纳，德国作曲家，尤以其浪漫歌剧著名，常以德国的传说为其作品基础）
World Fair 世界博览会
Gustave Eiffel 古斯塔夫·埃菲尔（1832–1923，法国工程师，为1889年的巴黎世界博览会设计埃菲尔铁塔）
Turk 突厥，土耳其人
Kiyomizu Temple 清水寺（日本京都）
Kyoto 京都（日本古都）
Easter Island 复活节岛（智利）
Chili 智利（南美洲西南部国家）
Polynesian 玻利尼西亚人（中太平洋岛屿原始居民）
Moai 摩艾（复活节岛巨像的别称）
Angkor Wat 吴哥窟（吴哥为柬埔寨西北的重要考古遗址）
Cambodia 柬埔寨（东南亚国家）
Suryavarman II 苏利阿伐曼二世（12世纪上叶高棉帝国的国王）
Khmer 高棉人（柬埔寨的一个民族，其文明在9世纪至15世纪时达到高峰）
Siem Reap 塞姆瑞菩村（柬埔寨）

Answer the following questions according to the text.

1. What is the significance of organizing the election of the New Seven Wonders of the World?
2. Can you name some of the New Seven Wonders candidates?
3. Why didn't these candidates win out?
4. Are you satisfied with the list of New Seven Wonders of the World and why?
5. What is your own list of the New Seven Wonders of the World and which wonder candidate do you prefer?

Unit 9 Architectural Functions

Warming Up

1. Do you know the basic structural methods used by architects?
2. Which will you give priority to when designing a building, utility or beauty?

TEXT A

A Synthesis of Form and Function

Buildings are where we live and work and spend much of our leisure time; of all the arts, it is architecture that forms the closest bond between us and our environment.

Architecture began as a response to the need to seek shelter from the elements and from wild beasts. The earliest shelters were not buildings but caves—but even here the occupants claimed their territories: 40,000 years ago the Australian Aborigines decorated the interior of their cave-dwellings with the oldest known expressions of art, and over 20,000 years later the people of Lascaux in present-day France ornamented their cave walls with amazingly sophisticated representations of their environment. Building techniques emerged many years later still, as a nomadic hunter-gatherer lifestyle gave way to a settled way of life based on agriculture.

Making the Most of Things

Until as recently as the 1850s, structural methods were essentially determined by the limits of building technology, the availability of materials, and regional conditions. In forested areas builders used timber; in rocky landscapes they used stone. Sometimes early builders cut directly into the solid rock of cliff faces, as in the 13th century BC temple of Ramesses at Abu Simbel in Egypt and the 1st AD treasury at Petra in Jordan. In places where there was neither timber nor rock, human ingenuity devised mud brick-molded blocks of sun-baked earth, the oldest manufactured building material that is still widely used today. The roof of Catalhoyuk, a Bronze Age settlement in Turkey from about 7000 BC, are of mud and brick, as is the core of the Ziggurat at Ur (c.2000 BC). And in our times the Great Mosque at Djenne in the West Africa is the world's largest mud brick building.

▲ Temple of Ramesses, Abu Simbel

▲ Remains of monumental building at Ur

▼ Great Mosque at Djenne

Two Basic Techniques

Post-and-beam, the earliest structural method, is still widely used today. A horizontal beam carries the weight of the floor above it and is supported by vertical posts. The Palace at Knossos (16th century BC) has masonry walls serving as posts, and timber beams; about 3,000 years later, China's Forbidden City, Japan's Castle of the White Heron, and Shakespeare's Globe Theatre in London employed timber posts and beams. Where permanence was important, early builders preferred stone. The elements could be rough-hewn and unornamented, like those of Stonehenge, or perfectly proportioned and exquisitely carved, like those of the Parthenon, built over 2,000 years later.

▲ Japan's Castle of the White Heron

But stone beams could not provide large spans between supports, which severely limited the enclosure of space, so the second important structural innovation was the arch, which can span considerable distances. Tapered blocks of stone, called voussoirs,

Pons Fabricius

are supported on a timber framework until the keystone is dropped into the crown of the arch. The keystone locks the structure into place and the framework can be removed. In the 1st century BC and the 1st century AD the Romans used this technique to build the Pons Fabricius and ceremonial Arch of Titus in Rome, and the Aqueduct in rance now known as the Pont du Gard.

Elaborations of the arch such as the vault (an extended arch) and the dome (an arch revolved) developed with the Roman invention of concrete at the height of the Roman Empire in the 1st and 2nd centuries AD. Clad in marble or other stone, concrete formed the structural core of buildings such as the Colosseum;or it could be left exposed, as in the dome of the Pantheon.

After the fall of Roman Empire the secret of concrete was lost and vaults were built with brick. In the 6th century, Byzantine architects extended the possibilities of dome construction with the invention of the pendentive—a triangular piece of masonry joining a four-piered square base to a circular dome. This brilliant innovation, which allowed greater freedom of plan and space, found its greatest expression in the church of Hagia Sophia in Istanbul. The exteriors of most Byzantine churches were quite plain, but the interiors glittered with gilt mosaics and colored marbles.

▲ Krak des Chevaliers in Syria

Early medieval buildings such as Krak des Chevaliers in Syria and the Tower of London, both built in the 11th century, were military in character with thick masonry walls pierced by small, defensible openings. Similar features characterize the walls and ceilings of Romanesque churches such as the 12th century cathedral of Santiago de Compostela in Spain.

VOCABULARY

New Words

synthesis *n.* 综和，合成

nomadic *a.* 游牧的，流浪的

Architectural Terms

brick-molded *a.* 砖模的

sun-baked earth 日光烘培土

voussoir *n.* 楔形拱石

keystone *n.* 拱顶石

rance *n.* 一种有蓝和白条纹的暗红色大理石（产于比利时）

Proper Names

Australian Aborigines 澳洲土著居民

Lascaux 拉斯考克斯（法国西南部一山洞）

Ramesses II 拉美西斯二世（古埃及法老）

Abu Simbel 阿布辛拜尔城（埃及）

Bronze Age 青铜器时代

Ziggurat 金子形神塔

Ur 乌尔城（古代美索不达米亚南部苏美尔的重要城市）

Great Mosque at Djenne 德贯尼大清真寺（西非）

Palace at Knossos 克诺索斯宫殿（希腊克里特岛）

Castle of the White Heron 白鹭城（日本姬路）

Shakespeare's Globe Theatre 莎翁环球大剧院

Pons Fabricius 伐布里修斯桥（古罗马）

Arch of Titus in Rome 提图斯凯旋门

Pont du Gard 伽合大桥（法国）

Krak des Chevaliers in Syria 十字军城堡（叙利亚）

Santiago de Compostela in Spain 圣地亚哥大教堂（西班牙）

EXERCISES

I. Match the English expressions with their Chinese equivalents.

1. brick-molded	A. 日光烘培土
2. sun-baked earth	B. 拱顶石
3. voussoir	C. 金字形神塔
4. keystone	D. 综合，合成
5. vault	E. 砖模的
6. synthesis	F. 澳洲土著居民
7. Ziggurat	G. 紫禁城（中国北京）
8. Australian Aborigines	H. 青铜器时代
9. Bronze Age	I. 拱顶
10. the Forbidden City	J. 楔形拱石

II. Decide whether the following statements are True or False.

1. Architecture began as a response to the need to seek shelter from the elements and from wild beasts.
2. Mud brick-molded blocks of sun-baked earth, the oldest manufactured building material, is never used today.
3. In our times the Great Mosque at Djenne in the West Africa is the world's largest mud brick building.
4. Early builders preferred timber if permanence of the buildings was strongly desired.
5. Arch cannot span as considerable distances as stone beams, which severely limited the enclosure of space.

III. Choose the best answer to each of the following questions.

1. According to the text, until as recently as the 1850s, structural methods were essentially determined by the limits of the following except ________.

   A. building technology B. the availability of materials

   C. regional conditions D. local laws

2. Post-and-beam is the earliest structural method in which a ________ carries the weight of the floor above it and is supported by ________.

   A. horizontal arch; vertical posts B. horizontal beam; vertical posts

   C. vertical post; horizontal arches D. vertical post; horizontal beams

3. Which of the following employs masonry walls serving as posts and timber beams?
   A. The Palace at Knossos.
   B. Japan's Castle of the White Heron.
   C. China's Forbidden City.
   D. Shakespeare's Globe Theatre in London.
4. Byzantine architects extended the possibilities of dome construction with the invention of the pendentive—a _______ piece of masonry joining a four-piered square base to a _______ dome.
   A. quadrangular; circular
   B. quadrangular; square
   C. triangular; circular
   D. triangular; square
5. A vault is a(n) _______ arch whereas a dome is a _______ arch.
   A. expanded; reversed
   B. expanded; revolved
   C. extended; revolved
   D. extended; reversed

IV. Oral task

In next class, you'll be asked to give an oral report based on one of the following questions. Work in teams and search the library or Internet for relevant pictures, facts or stories to support your points.

→ 1 Try to illustrate the two basic structural methods.

→ 2 What is "the pendent" and in what way did it change the dome construction?

→ 3 Why are buildings called "a synthesis of form and function"?

TEXT B

Castles, Palaces, and Forts

Whether built primarily for defense or for show, castles, palaces, and forts have always expressed the power and wealth of their owners.

Nowadays we think of castles, palaces, and forts as synonymous, but these buildings originally had separate functions. The word "palace" is derived from the Latin *palatium*, which was the name of the hill in Rome where the Emperor Domitian built his private residence at the end of the 1st century AD. Its official name was Domus Augustana, but it soon became known as the Palatium after the hill itself. Architectural splendor went hand-in-hand here with connotations of power and authority; hence the emergence in later centuries of the tradition of giving the name "palatium" to any substantial dwelling occupied by a ruler or a group of rulers. Hadrian's Villa at Tivoli is an early case in point.

▲ Domus Augustana

"Castle" is derived from the Latin *castrum*, and its diminutive form *castellum*—originally meaning a Roman military camp, but later, by the middle of the 11th century, fortified residence occupied by a king or baron. It is the residential aspect of the castle and the exclusiveness of its ownership that distinguishes it from other type of fortress. When the need for fortified residences no longer exists or when a central authority is strong enough to curtail their construction, as in France and England during the 16th century, the history of the castle as a viable institution is over and the great house or palace, unfortified, replaces it as the characteristic residence of the ruling class.

▲ Hadrian's Villa at Tivoli

However, this distinction between palaces and castles is not always so readily apparent. For example, in countries where relative peace has been achieved after a period of military or civil strife, palaces are likely to restrain some of the defenses associated with castles. Conversely, castle that is still used as residences long after their military importance has ceased may be converted into palaces. Most major civilizations since the third millennium BC have built palaces, but in ways that are so diverse in planning, construction, and decoration as to defy easy classification. From a purely formal point of view, however, it is possible to detect two main lines of development: the first rooted in the pavilion-like palaces of the Eastern world, the second in the block-like palaces of the Western world.

Changing Interpretations

At certain times and in certain places these two lines of development have overlapped or coalesced, but most historians would agree that whereas in the East the tendency has been a favor delicacy of scale over overt monumentality, in the West the reverse has been true. While it is known that certain palaces in the East were built without the protection of an outer defensive wall, they are comparatively rare. But in the West, from the 16th century onward, the unfortified palace rapidly became the rule rather than the exception. The 18th century Baroque Residenz at Wuzburg is an archetypal example of this style.

▼ Baroque Residenz at Wuzburg

Nowadays, we tend to use the word "castle" and "palace" as metaphors for grandeur, opulence, luxuriousness of appointments, and defensive capability. William Randolph Hearst's San Simeon estate in California possesses all of these qualities—hence the durability of that other name that was bestowed on it by visitors in the 1920s: Hearst Castle.

▲ Hearst Castle

The Great Wall of China

Homes, towns, cities, military outposts, and national frontiers have been fortified against enemy invasion throughout human history—but never on a scale as vast as the Great Wall built along the northern frontier of China between the 3rd century BC and the 17th century AD.

There are, or were, four Great Walls

of China, all intended to protect the country from the most aggressive of its adversaries—the nomadic people of Mongolia to the north. The first wall was built between 217 and 208 BC by the First Emperor of Qin. Convicts and hundreds of thousands of peasants worked on the project under Qin's harsh scheme of labor conscription. This wall, approximately 1,800 miles (almost 3,000 km) long, ran from Shanhaiguan in the east to Yemenguan in the west and was punctuated by as many as 25, 000 watch towers.

The second Great Wall was founded in the early years of the 1st century BC by Emperor Wu of the Han Dynasty, who pushed the first wall westward for another 300 miles (500 km). The third Great Wall was also an extension, of the similar length to the second wall but it ran in the opposite direction. Built between 1138 and 1198, it led from the old eastern terminus at Shanhaiguan and snaked its way north to Dandong on the Yalu River.

The best preserved and most formidable wall, and the one most commonly known as the Great Wall of China, is the fourth. Work on this relatively new structure started in 1368 during the reign of the First Emperor of the Ming Dynasty, Zhu Yuanzhang, and continued until the fall of the last Ming emperor in 1644. By this stage the wall, with its various offshoots, had reached an extraordinary length of about 4,000 miles (6,500 km).

VOCABULARY

New Words

diminutive *a.* 特小的
baron *n.* 男爵
curtail *v.* 缩小
viable *a.* 切实可行的
strife *n.* 冲突
coalesce *v.* 联合，合并
overt *a.* 公开的
archetypal *a.* 原型的，典范的
conscription *n.* 征兵，招兵
punctuate *v.* 打断
terminus *n.* 终点，边界
formidable *a.* 强大的，令人生畏的
offshoot *n.* 枝条，分枝

Architectural Terms

fort *n.* 堡垒，要塞

watch tower 瞭望塔（烽火台）

Proper Names

Domitian 图密善(古罗马皇帝, 81–96年在位)
Domus Augustana 奥古斯塔纳圆顶殿（意大利）
Hadrian's Villa at Tivoli 古罗马哈德良皇帝宅邸（意大利）
Residenz at Wuzburg 坐落在伍兹伯格的宅邸
William Randolph Hearst 威廉·伦道夫·赫斯特（美国报刊和杂志出版商）
Hearst Castle 赫斯特堡（威廉·伦道夫·赫斯特在美国加利福尼亚圣西蒙为自己修建的富丽堂皇的庄园）
Shanhaiguan 山海关
Yemenguan 雁门关
Dandong 丹东
the Yalu River 鸭绿江

EXERCISES

I. Translate the following English into Chinese and Chinese into English.

1. fort ______________
2. fortress ______________
3. pavilion ______________
4. grandeur ______________
5. splendor ______________
6. 瞭望塔 ______________
7. 富丽 ______________
8. 终点站 ______________
9. 枝条, 分枝 ______________
10. 重叠 ______________

II. Complete the following table according to the text.

	Time	Emperor	Length	Running from ____ to ____
The first Great Wall				
The second Great Wall				
The third Great Wall				
The fourth Great Wall				

III. Oral task

In next class, you'll be asked to give an oral report based on one of the following questions. Work in teams and search the library or Internet for relevant pictures, facts or stories to support your points.

1. How do you understand "castles, palaces, and forts have always expressed the power and wealth of their owners"?
2. Do you think Hearst Castle performed its intended function?
3. Do you think the Great Wall of China serves as a good fort since it was founded?

TEXT C

Safety Factors in Designing a Building

Design of buildings for both normal and emergency conditions should always incorporate a safety factor against failure. The magnitude of the safety factor should be selected in accordance with the importance of a building, the extent of personal injury or property loss that may result if a failure occurs, and the degree of uncertainty as to the magnitude or nature of loads and the properties and behavior of building components.

Safety factors for various building system are discussed here. This section presents general design principles for protection of buildings and occupants against high winds, earthquakes, water, fire, and lightning.

Wind Protection

For practical design, wind and earthquake may be treated as horizontal, or lateral, loads. Although wind and seismic loads may have vertical components, these generally are small and readily resisted by columns and bearing walls.

The variation with height of the magnitude of a wind load for a multi-storey building differs from that of seismic load. Nevertheless, provisions for resisting either type of load are similar.

In areas where the probability of either a strong earthquake or a high wind is small, it is nevertheless advisable to provide in buildings considerable resistance to both types of load. In many cases, such resistance can be incorporated with little or no increase in costs over designs that ignore either high wind or seismic resistance.

Protection Against Earthquakes

Building should be designed to withstand minor earthquakes without damage, because they may occur almost everywhere. For major earthquakes, it may not be economical to prevent all damages but collapse should be preluded.

The principles require that collapse be avoided, oscillations of building damped, and damage to both structural and nonstructural components minimized. Nonstructural components are especially liable to damage from large drift. For example, walls are likely to be stiffer than structural framing and therefore subject to greater seismic forces. The wall as a result may crack or collapse. Also they may interfere with planned action of structural components and cause additional damage. Consequently, a seismic design of buildings should make allowance for large drift, for example, by providing gaps between adjoining building components not required to be rigidly connected together and by permitting sliding of such components. Thus, partitions and windows should be free to move in their frames so that no damage will occur when an earthquake wrecks the frames.

Protection Against Water

Whether thrust against and into a building by a flood, driven into the interior by a heavy rain, leaking from pluming, or seeping through the exterior enclosure, water can cause costly damage to a building. Consequently, designers should protect buildings and their contents against water damage.

Protective measures may be divided into two classes: floodproofing and waterproofing. Floodproofing provides protection against flowing surface water, commonly caused by a river overflowing its banks. Waterproofing provides protection against penetrating through the exterior enclosure of buildings of groundwater, rainwater, and melting snow.

Protection Against Fire

There are two distinct aspects of fire protection: life safety and property protection. Although providing for one aspect generally results in some protection for the other, the two goals are not mutually inclusive. A program that provides for prompt notification and evacuation of occupants meets the objectives for life safety, but provides no protection for property. Conversely, it is possible that adequate property protection might not be sufficient for protection of life.

Absolute safety from fire is attainable. It is not possible to eliminate all combustible materials or all potential ignition sources. Thus, in most cases, an adequate fire protection plan must assume that unwanted fires will occur despite the best efforts to prevent them. Means must be provided to minimize the losses caused by the fires that do occur.

The first obligation of designers is to meet legal requirements while providing the facilities required by the client. In particular, the requirements of the applicable building code must be met. The building code will contain fire safety requirements, or it will specify some recognized standard by reference. Many owners will also require that their own insurance carrier be consulted—to obtain the most favorable insurance rate, if for no other reasons.

Lightning Protection

Lightning, a high-voltage, high-current electrical discharge between clouds and the ground, may strike and destroy life and property anywhere thunderstorms have occurred in the past. Buildings and their occupants, however, can be protected against this hazard by installation of a special electrical system. Because an incomplete or poor installation can cause worse damage or injuries than no protection at all, a lightning-protection system should be designed and installed by experts.

As an addition to other electrical systems required for a building, a lightning-protection system increases the construction cost of a building. A building owner therefore has to decide whether potential losses justify the added expenditure. In doing so, the owner should take into account the importance of the building, danger to occupants, value and nature to building contents, type of construction, proximity of other structures or trees, type of terrain, height of building, number of days per year during which thunderstorms may occur, cost of disruption of business or other activities and the effects of loss of essential services, such as electrical and communication system. (Buildings housing flammable or explosive materials generally should have lightning protection.) Also, the owner should compare the cost of insurance to cover losses with the cost of the protection system.

VOCABULARY

New Words

seismic *a.* 地震的
prelude *v.* 拉开序幕
oscillation *n.* 摆动，动摇
seep *v.* 渗漏
evacuation *n.* 疏散，撤离
combustible *a.* 可燃的，易燃的
ignition *n.* 着火，点火
disruption *n.* 分裂

Architectural Terms

bearing wall 承重墙
multi-storey building 多层建筑

Answer the following questions according to the text.

1. Why could the wind and earthquake be treated as horizontal not vertical?
2. How could the building be protected against water damages?
3. What are the two distinct aspects of fire protection?
4. What should a building owner take into consideration before installing the lightning-protection system?
5. Suppose you are an architect, what specific suggestions will you give to your clients as to how to balance safety and economy factors?

10 Unit Building Materials

Warming Up

1. With your partner, list out the major building materials used by builders.
2. Work in groups to compare two major building materials—wood and brick. Make a list of the advantages and disadvantages of the two.

TEXT A

Building Materials in Britain

The principal materials which have been used in Britain for building structures of particular architectural interest are: brickwork, clapboard, concrete, cruck building, facing, glass, masonry, plasterwork, roof, slate, stone, tiles, timber-framing, walling. Other materials, which have been widely used, were employed in primitive times, or for smaller, everyday dwellings, or were traditional for building in certain parts of the country where other materials, such as stone or wood, were not readily available in quantity. Most important among these are the earths, wood, thatch, flint and pebbles, which are discussed here.

Unburnt clay earth, necessarily mixed with some other substances, has been a traditional building material in Britain since very early times. By the 16th and 17th centuries its use was mainly confined to rural structures—cottages and small farmhouses—but in Norman times many town houses had mud walls. There are several ways of using earth for building and these methods have been developed in different parts of the country (and under various names) because of the local content of the material to be found there. Construction using earths can be of the plastic type where a clay earth is mixed with water to give a stiff mud consistency or it can be of dry earth rammed to give cohesion. Clay earth mixed with water to give a stiff mud must contain lime to make it set, also straw, gravel and sand. The gravel may be small pebbles or fragments of slate or similar aggregate. Common names for such "mixes" are cob (mainly used in the south-west of the country), clob (in Cornwall) or mud.

The most usual roofing material traditionally used for buildings made from these various types of clay earth, as well as those of half-timber construction, was some form of thatch. This was because it was the most lightweight of roofing means and therefore suitable for walls which would not bear a great load. Throughout northern Europe buildings had been roofed since earliest times with turves, moss, heather or reeds and, often, a mixture of more than one of these. From Norman times the most usual thatching materials had been reed, straw or heather, the last being especially used in moorland areas where the material was plentiful, such as Scotland, northern England and the West Country. The chief problem with such roofing materials was fire danger and, because of this, restrictions on thatching in towns came into force as early as the 13th century, and a plaster covering was often added to such roofs to attempt to counter this hazard. In rural areas thatch as a roofing material has survived to the present day. The thatcher's art has been developed over the centuries and many roofs are decoratively cut and trimmed especially along the ridge where reinforcement is needed.

▲ Church of St. Andrew at Greensted

While earth and clay made into mud were the earliest materials for building in Britain, in general use before the coming of the Romans, wood was also utilized before the 1st century BC, but for the more important structures. Almost nothing survives of building made of wood before the Middle Ages, when timber-framed structure was developed. One interesting example is the Church of St. Andrew at Greensted in Essex. In the 10th century a small wooden chapel was built here with walls composed of trunks falling outwards and were let into a sill at the base and a plate at the top; they were fasten with wood pins. These timbers still form the basis of the nave wall of the present church though they had to be removed in the 19th century, shortened and set into a brick plinth because of decay at the foot. This type of construction, profligate in its use of oak trunks, was typical for centuries in areas of Europe where wood was plentiful, as in Norway and the forest areas of Rumania. For a doubtlessly similar reason, it was not uncommon in Saxon England, but it is quite different from the later English construction of timber-framing.

Flint is an extremely durable material which has been used for building in England from the Iron Age onwards. It is a pure form of silica, extremely hard but easy to split. Flint occurs in chalk deposits and is irregularly shaped nodules, white on the exterior from their long sojourn in the chalk, but with shades of dark grey or black inside. The use of flint has been traditional in chalk areas in south-east and southern England as far west as Wiltshire and Dorset. The finest-quality work, architecturally and decoratively, is in East Anglia where, in the Middle Ages, flint was a predominant building material for churches.

Cobbles are pieces of rock which have been broken away by the action of glaciers or water and have been collected and used for rubble wall building or facing since early times. Cobbles vary greatly in size from three or four inches in diameter to a foot or more. Pebbles, which are smaller, are particularly in use in sea-coast areas.

VOCABULARY

New Words

thatch *n.* 茅草
flint *n.* 燧石，火石
ram *v.* 撞击，夯入
aggregate *n.* 集料，骨料
turf *n.* 草皮
heather *n.* 欧石南（杜鹃花科植物）
moorland *n.* 高沼地
profligate *a.* 恣意挥霍的
nodule *n.* 小圆块
sojourn *n.* 逗留

Architectural Terms

clapboard *n.* 墙板，隔板
cruck building 曲木屋架建筑
facing *n.* （起保护作用的）面层，饰面
plasterwork *n.* 灰泥工艺
slate *n.* 石板，石片瓦
nave *n.* 中堂，正厅，正殿
plinth *n.* 底座，柱基
silica *n.* 硅石，硅土

Proper Names

Norman 诺曼人（10世纪定居于诺曼底的斯堪的纳维亚人和法国人的后裔）
Church of St. Andrew 圣安德鲁大教堂（英国埃塞克斯郡）
Essex 埃塞克斯郡（英格兰东北部）
Rumania 罗马尼亚
Wiltshire 威尔特郡（英格兰南部）
Dorset 多塞特郡（英格兰南部）

EXERCISES

I. Match the English expressions with their Chinese equivalents.

1. clapboard	A. 灰泥工艺, 石膏工艺
2. cruck building	B. 脊部
3. facing	C. 墙板, 隔板
4. plasterwork	D. 中堂, 正厅, 正殿
5. slate	E. 硅石, 硅土
6. ridge	F. 小教堂, 礼拜堂
7. chapel	G. 石板, 石片瓦
8. nave	H. 覆面
9. plinth	I. 曲木屋架建筑
10. silica	J. 底座, 柱基

II. Decide whether the following statements are True or False.

1. Burnt clay earth, necessarily mixed with some other substances, has been a traditional building material in Britain since very early times.
2. There are several ways of using earth for building and these methods have been developed in different parts of the Britain because of the local content of the material to be found there.
3. Because of fire danger, restrictions on thatching in towns came into force as early as the 13th century.
4. In rural areas thatch as a roofing material hasn't survived to the present day.
5. Varying greatly in size from three or four inches in diameter to a foot or more, cobbles are smaller than pebbles.

III. Choose the best answer to each of the following questions.

1. Many building materials, excluding ________, were readily available in quantity in Britain at ancient times.
   A. brick B. tile C. stone and wood D. plaster
2. Because _________ was the most lightweight of roofing means, it is suitable for walls which would not bear a great load.
   A. clay B. thatch C. timber D. mud
3. The thatcher's art has been developed over the centuries and many roofs are decoratively cut and trimmed especially along the ________ where reinforcement is needed.
   A. roof B. eaves C. ridge D. beam

4. ________ made into ________ were the earliest materials for building in Britain, in general use before the coming of the Romans.
   A. Earth and clay; mud B. Earth and mud; clay
   C. Mud; earth and clay D. Clay; earth and mud
5. ________ is an extremely durable material which has been used for building in England from the Iron Age onwards.
   A. Clay B. Gravel C. Timber D. Flint

IV. Oral task

In next class, you'll be asked to give an oral report based on one of the following questions. Work in teams and search the library or Internet for relevant pictures, facts or stories to support your points.

→ 1 What do you know about the building materials before you read the text?

→ 2 What do you learn about the building materials from the text?

→ 3 How can cobbles be used in buildings in your opinion?

TEXT B

Brickwork in Britain

A brickwork is a block of regular form moulded and cut from a soft material such as clay, then hardened, its particles consolidated by fusion, under heat from the sun or being burnt in a kiln. The color and texture varies according to the clay in use, the method of burning and the process of manufacture.

Brick has been widely used as a building material for over 6,000 years. It has many useful qualities. It is inexpensive to produce and the raw material is readily to hand. It can be made in different sizes, colors, qualities and textures. It is durable, withstands the weather and is a good insulating material. It can be used in conjunction with other materials as, for example, in Roman walling, where the bricks provided bonding courses at intervals between others of flint or rubble masonry, and in decorative form as infilling to timber framing or with stone dressing—a treatment often employed by Sir Christopher Wren (Hampton Court Palace, 1689-1701).

Early brick-making in areas of the world where the climate was hot, such as Egypt and Mesopotamia, used the sun to dry and heat the material. The Spanish word *adobe*, meaning a sun-dried brick, reflects a method employed until modern times and such brickwork can be seen in many parts of the world including Spain, south-east Europe and South America. Burnt bricks were being made in the Near East by 3000 BC and, in Britain, some examples have been found which predate the Roman occupation. These bricks are not very hard and certainly softer than Roman ones.

The Romans widely employed bricks for building, especially in provincial works. This was the case in Britain, where brickwork was practiced extensively. The bricks were used as bonding courses in walling, as can be seen in remains of town walls such as those at Cochester in Essex, and also facing to a concrete core. Roman building bricks were large and thin, resembling large tiles and the mortar courses between were of a thickness generally about the same as the bricks themselves. The bricks, which were hard and well-burnt, varied in size from one inch to one-and-a-half inches in thickness and could be square or rectangular, the latter ranging from 18in×12in to 12in×6in. Floor bricks were rectangular but smaller and thicker and those made to build up into pilae (underfloor piers for the heating system) were square. Hollow box bricks were made for insertion into walls for cavity heating methods.

After the Romans left, brick building died out in Britain and, unlike northern Europe in the Hanseatic Towns along the Baltic coast, and unlike France and Italy, where brick was widely used from the 12th century onwards, the crafts were slowly to re-appear. There are very few instances from about 1200 of the use of brick dressing in Britain but the first example where a considerable number of bricks were used, in conjunction with other materials was Little Wenham Hall in Suffolk (c. 1275). It is likely, though this is not documented, that these bricks of varied hues were made on the site. Brick continued only in limited use as a building material in Britain until the 17th century. Holy Trinity Church in Hull is one of the few 14th-century examples and even in the 15th century, when the material was more widely used, chiefly for cottages and small houses. Brick was, however, beginning to be found useful in areas where stone was not available, for the great forest covering of the country was at last showing signs of depredation.

▲ Holy Trinity Church in Hull

In Tudor period, brick building developed more quickly. During the time of Henry VIII, the work was more skillfully handled, both in the making of bricks and in the laying of them. Decoration was attempted on an ambitious scale and in a manner which had been traditional in the Hanseatic towns of Holland, Germany or Poland since the 13th century. Diaper patterns were produced by using headers with darker, more heavily burnt ends. The picturesque skylines of the time with their numerous chimney-stacks provided the opportunity for the greatest variety of ornamentation in brick. Notable examples include Hampton Court Palace, Norham Hall, Essex, Compton Wynyates, Warwickshire and Hengrave Hall, Suffolk.

▲ Compton Wynyate

▼ Hengrave Hall

A further advance in the handling of brick was apparent in the 1630s and 1640s. Classical architecture was beginning to be accepted and, particularly in houses, the varied forms of decorative gable were fashionable. Both these architectural factors provided a challenge to the brick builders since the material is more difficult to handle in the precise form required than stone which can be more easily carved.

The century which commenced in 1660 was the great age of English brick building. High-quality bricks were made from a variety of fine clays, and were beautifully laid, and used to decorate all kinds of buildings. Eighteenth-century work showed a taste for polychromy in soft reds, greys and yellow. Brick building became much less costly as bricks were manufactured on a large scale so, gradually, towns replaced half-timber buildings with fire-proof brick ones.

Despite the increasing use of the modern materials of concrete and steel, brick is still a major building material in Britain and about two percent of the product is hand-made.

VOCABULARY

New Words

insulate *v.* 使绝缘

infilling *n.* 填入，填入之物

cavity *n.* 腔，洞

Architectural Terms

kiln *n.* 窑

rubble *n.* 碎石

adobe *n.* 砖坯

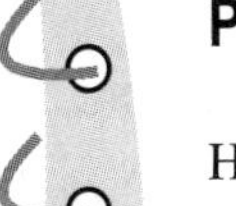

Proper Names

Hampton Court Palace 汉普敦王宫（英国）

Hanseatic Towns （中世纪）商业同业公会的市镇

Baltic coast 波罗的海沿岸

Little Wenham Hall in Suffolk 萨福克郡小温海姆宅邸（英国）

Holy Trinity Church in Hull 赫尔市圣三一教堂（英国）

Norham Hall Essex 诺汉姆宫（英国埃塞克斯郡）

Compton Wynyates, Warwickshire 康普顿宅邸（英国沃里克郡）

Hengrave Hall, Suffolk 亨格瑞夫宅邸（英国萨福克郡）

EXERCISES

I. Translate the following English into Chinese and Chinese into English.

1. kiln ________________
2. adobe ________________
3. mortar ________________
4. rubble ________________
5. consolidate ________________
6. 熔化，合成 ________________
7. 使绝缘 ________________
8. 填入，填入之物 ________________
9. 长方形的 ________________
10. 颜色，色彩 ________________

II. Complete the following flowchart.

Brick Making Process

Step One → Step Two → Step Three → Step Four

III. Oral task

In next class, you'll be asked to give an oral report based on one of the following questions. Work in teams and search the library or Internet for relevant pictures, facts or stories to support your points.

→ 1 What are the useful qualities of brickwork?

→ 2 What's the difference between early brick making and modern brick making?

→ 3 What had happened to brickwork in the 18th century?

TEXT C

Wood Construction

Wood is a plentiful and economic building material that comes from a renewable resource—trees, which are often planted with a purpose and then harvested with a particular end product in mind.

As a consequence of its origin, wood as a building material has inherent characteristics with which users should be familiar. For example, although cut simultaneously from trees growing side by side in a forest, two boards of the same species and size most likely do not have the same strength. The task of describing this nonhomogeneous material, with its variable biological nature, is not easy. But it can be described accurately, and much better than was possible in the past, for research has provided much useful information on wood properties and behavior in structures.

Research has shown, for example, that a compression grade cannot be used, without modification, for the tension side of a deep bending member. Also, a bending grade cannot be used, unless modified, for the tension side of a deep bending member, or for a rimental to tensile strength than to compressive strength. Furthermore, research has made possible better estimates of engineering qualities of wood. No longer is it necessary to use only visual inspection, keyed to averages, for estimating the engineering qualities of a piece of wood.

Practice, not engineering design, has been the criterion in the past in home building. As a result, the strength of wood often was not fully utilized. With a better understanding of wood now possible, the availability of sound structural design criteria, and development of economical manufacturing processes, greater and more efficient use is being made of wood in home building.

Improvements in adhesives also have contributed to the betterment of wood construction. In particular, the laminating process, employing adhesives to build up thin boards into deep timbers, improves on nature. Not only are stronger structural members thus made available, but also higher grades of lumber can be placed in regions of greatest stress, and lower grades in regions of lower stress, for overall economy. Despite variations in strength of wood, lumber can be transformed into glued-laminated timbers of predicable strength and with very little variability in strength.

Basic Characteristics and How to Use Them

Wood differs in several significant ways

from other building materials. Its cellular structure is responsible, to considerable degree, for this. While most structural materials are essentially isotropic, with nearly equal properties in all directions, wood has three principal grain directions—longitudinal, radial, and tangential. (Loading in the longitudinal direction is referred to as parallel to the grain, whereas transverse loading is considered across the grain.) Parallel to the grain, wood possesses high strength and stiffness. Across the grain, strength is much lower. (In tension, wood stressed parallel to the grain is 25 to 40 times stronger than when loaded perpendicular to the grain.) Furthermore, a wood member has three moduli of elasticity, with a ratio of largest to smallest as large as 150:1.

Wood undergoes dimensional changes from causes different from those dimensional changes in most other structural materials. For instance, thermal expansion of wood is so small as to be unimportant in ordinary usages. Significant dimensional changes, however, occur because of grain or loss in moisture. Swelling and shrinkage from this cause vary in the three grain directions; size changes about 6 to 16% tangentially, 3 to 7% radially, but only 0.1 to 0.3% longitudinally.

Wood offers numerous advantages nevertheless in construction applications beauty, versatility, durability, workability, low cost per pound, high strength-to-weight ratio, good electrical insulation, low thermal conductance, and excellent strength at low temperatures. It is resistant to many chemicals that are highly corrosive to other materials. It has high shock-absorption capability. It can withstand large overloads of short time to sharp curvature. A wide range of finishes can be applied for decorative purposes. Wood can be used in both wet and dry applications.

In addition, a wide variety of wood framing systems is available. The intended use of a structure, geographical location, configuration required, cost, and many other factors determine the framing system to be used for a particular project.

VOCABULARY

New Words

laminated *a.* 压层的
isotropic *a.* 等向的
longitudinal *a.* 纵向的
tangential *a.* 正切的
perpendicular *a.* 成直角的，垂直的
moduli *n.* (modulus 的复数) 模具，系数
elasticity *n.* 弹性，活力
versatility *n.* 多面性
conductance *n.* 传导
corrosive *a.* 腐蚀性的
curvature *n.* 弯曲，弯曲的形状

Answer the following questions according to the text.

1. In what way is wood special among the building materials?
2. Why is wood called "nonhomogeneous material"?
3. What was the criterion in home building in the past?
4. What are the three principal grain directions of wood?
5. What are the advantages of wood as a kind of building material?

Appendix I: Keys to Exercises

Unit 1

TEXT A

I. 1-E 2-C 3-G 4-D 5-A 6-J 7-F 8-B 9-I 10-H

II. 1-T 2-F 3-T 4-T 5-F 6-F

III. 1-B 2-C 3-C 4-D 5-A

TEXT B

I.

1. 节能建筑
2. 测量员，检查员
3. 生态可持续型城市
4. 建筑管理官员
5. 地方权威规划部门
6. circle template
7. hierarchy
8. apprenticeship
9. proprietor
10. artifact

II.

1. architects scale
2. adjustable triangle
3. compass
4. computer
5. turquoise pencil
6. circle template
7. French curve

Unit 2

TEXT A

I. 1-B 2-D 3-F 4-E 5-H 6-I 7-J 8-G 9-A 10-C

II. 1-T 2-F 3-F 4-T 5-F 6-T

III. 1-C 2-D 3-C 4-B 5-A

TEXT B

I. 1. 大理石 2. 清真寺 3. 比例 4. 釉面砖 5. 基本图案
6. entablature 7. prevail 8. Rococo style 9. neoclassicism 10. pediment

II. 1-T 2-F 3-T 4-F 5-F 6-T

III. 1-B 2-D 3-C 4-A 5-D

Unit 3

TEXT A

I. 1-C 2-G 3-I 4-J 5-D 6-A 7-H 8-B 9-F 10-E

II. 1-T 2-F 3-T 4-F 5-T

III. 1-B 2-D 3-C 4-A 5-C

TEXT B

I. 1. 新古典主义风格 2. 门廊，柱廊 3. 挖掘 4. 蚀刻版画 5. 几何形状
6. tapestry 7. ceramics 8. architectural ethos 9. rotunda 10. antiquity

II. 1-D 2-C 3-A 4-G 5-B 6-H 7-E 8-I 9-F

Unit 4

TEXT A

I. 1-F 2-I 3-H 4-A 5-D 6-G 7-J 8-B 9-E 10-C

II. 1-F 2-T 3-F 4-T 5-T

III. 1-B 2-C 3-D 4-A 5-D

TEXT B

I. 1. 房地产 2. 卢浮宫金字塔 3. 玻璃覆盖的 4. 几何图形式的设计
5. 本地的；本土的 6. urban design 7. proviso 8. partnership
9. municipal building 10. graft

II. 1-c 2-e 3-b 4-g 5-j 6-a 7-l 8-h 9-d 10-k 11-f 12-i 13-n 14- m

Unit 5

TEXT A

I. 1-D 2-B 3-G 4-A 5-J 6-E 7-I 8-F 9-C 10-H

II. 1-F 2-T 3-F 4-F 5-T

III. 1-D 2-A 3-B 4-C 5-B

TEXT B

I. 1. 如悬臂向外伸展 2. 平台, 台地 3. 华盖, 天篷 4. 平板, 平板状物
5. 壁架, 窗台 6. 场所, 地点 7. 指示牌 8. 一分为二
9. 有机主义风格的 10. 住所

II. 1. disrupted 2. sculpture 3. natural site 4. cantilevered 5. human
6. stability; permanence; ephemeralness; paradox

Unit 6

TEXT A

I. 1-D 2-J 3-F 4-I 5-G 6-E 7-H 8-A 9-B 10-C

II. 1-F 2-T 3-F 4-F 5-T 6-T

III. 1-C 2-D 3-A 4-B 5-D 6-C

TEXT B

I. 1. 小亭子 2. 石碑 3. 牌匾, 匾额 4. 天窗, 老虎窗 5. 观景楼
6. ridge 7. veranda 8. balustrade 9. stupa 10. spire

II. 1-T 2-F 3-T 4-F 5-T 6-F

III. 1-A 2-B 3-D 4-D 5-B

Unit 7

TEXT A

I. 1-D 2-G 3-B 4-I 5-A 6-F 7-J 8-C 9-H 10-E

II. 1-F 2-T 3-F 4-T 5-T

III. 1-B 2-C 3-C 4-A 5-D

TEXT B

I. 1. 古典主义，古典风格 2. 柱廊 3. 拱门 4. 建筑物的正面 5. 市政厅
6. 檐口 7. 角楼, 塔楼 8. 中世纪建筑 9. 棋盘式街道布局 10. 锥形屋顶

II. a- 9 b-7 c-1 d-5 e-3 f-2

Unit 8

TEXT A

I. 1-F 2-I 3-A 4-J 5-C 6-B 7-G 8-E 9-H 10-D

II. 1-F 2-T 3-F 4-T 5-F

III. 1-C 2-D 3-C 4-A 5-B 6-D

TEXT B

I. 1. 平面设计, 布局 2. 原型 3. 古罗马圆形剧场 4. 圆形露天剧场 5. 竞技场
6. facade 7. worshipper 8. mortar 9. Art Deco-style 10. fortification

II. 1-T 2-T 3-F 4-F 5-F

III. 1-C 2-A 3-B 4-C 5-D

Unit 9

TEXT A

I. 1-E 2-A 3-J 4-B 5-I 6-D 7-C 8-F 9-H 10-G

II. 1-T 2-F 3-T 4-F 5-F

III. 1-D 2-B 3-A 4-C 5-C

TEXT B

I. 1. 堡垒, 要塞 2. 碉堡, 堡垒, 要塞 3. 亭子 4. 伟大, 宏大 5. 壮丽, 壮观
6. watch tower 7. opulence 8. terminus 9. offshoot 10. overlap

II.

	Time	Emperor	Length	running from ___ to ___
The first Great Wall	between 217 and 208 BC	the First Emperor of Qin	1,800 miles	from Shanhaiguan in the east to Yemenguan in the west
The second Great Wall	in the early years of the 1st century BC	Emperor Wu of the Han Dynasty	pushed the first wall westward for another 300 miles	
The third Great Wall	between 1138 and 1198		the similar length to the second wall but it ran in the opposite direction	from the old eastern terminus at Shanhaiguan and snaked its way north to Dandong on the Yalu River
The fourth Great Wall	from 1368 to 1644	the First Emperor of the Ming Dynasty, Zhu Yuanzhang; and continued more or less until the last Ming emperor		

Unit 10

TEXT A

I. 1-C 2-I 3-H 4-A 5-G 6-B 7-F 8-D 9-J 10-E

II. 1-F 2-T 3-T 4-F 5-F

III. 1-C 2-B 3-C 4-A 5-D

TEXT B

I. 1. 窑 2. 砖坯 3. 砂浆, 灰浆 4. 碎石 5. 巩固, 加强
6. fusion 7. insulate 8. infilling 9. rectangular 10. hue

II.

Step One mould and cut from a soft material such as clay

Step Two harden

Step Three consolidate by fusion

Step Four heat under the sun or burn in a kiln

Appendix II : Glossary

New Words

A

abbe *n.* 僧侣, 道士 U6C

abundance *n.* 充裕, 丰富 U3C

academic *a.* 学术上的 U2C

acclaim *n.* 欢呼, 称赞 U5B

acoustics *n.* (房间, 剧场等的) 音质, 声学 U4C

adhere *v.* 粘附, 附着 U1A

adjoin *v.* 贴近, 邻接 U4C

adornment *n.* 装饰, 装饰物 U6B

aesthetic *a.* 美学的, 艺术的 U1A

affirm *v.* 证实 U3A

aggregate *n.* 集料, 骨料 U10A

allegorical *a.* 寓言的, 讽喻的 U2C

aluminum *n.* 铝 U7C

amidst *prep.* (=amid) 在……当中, 在……中间 U5C

amphitheater *n.* (古罗马) 圆形的露天剧场 U4C

amplification *n.* 扩音 U4C

anatomist *n.* 解剖学家 U4C

antiquity *n.* 古迹, 古物 U3B

antisepsis *n.* 防腐 (法), 抗菌 (法) U6A

apprenticeship *n.* 学徒身份, 学徒年限 U1B

Arabic calligraphy 阿拉伯字体书写 U2B

archaeologist *n.* 考古学家 U3B

archetypal *a.* 典型的, 原型的 U2A

articulated *a.* 具有表现力的 U7C

artifact *n.* 人工制品 U1B

ascetic *n.* 修道者 U6C

aspiring *a.* 有志气的, 有抱负的 U4A

assimilate *v.* 同化, 吸收 U7C

assimilation *n.* 吸收, 同化 U1A

assorted *a.* 多样的, 混合的 U6C

astronomy *n.* 天文学 U4C

attest to 证明, 表明 U3A

aura *n.* 气味, 气氛, 氛围 U5C

austere *a.* 朴素的, 无装饰的 U4C

awesome *a.* 令人敬畏的 U2A

axis *n.* 轴, 轴线 U6A

B

backdrop *n.* 幕布, 背景 U5C

banality *n.* 平庸, 陈腐 U7A

baron *n.* 男爵 U9B

beacon *n.* 灯塔, 烽火信号 U6B

beacon chamber (灯塔) 照明室, 信号室 U8A

beam *n.* 光束 U8A

beneficence *n.* 德行, 善行, 仁慈 U6C

bestow *v.* 给予, 赠与 U4B

bishop *n.* 主教 U4C

bold *a.* 粗线条的 U2A

Buddha *n.* 佛, 佛陀 U6C

Buddhist *a.* 佛教的 U6A

C

carnival *n.* 狂欢, 欢聚 U8C

cavity *n.* 腔, 洞 U10B

celestial *a.* 天空的, 天国的 U6C

centennial *n.* 一百周年纪念 U8C

civic *a.* 城市的，市民的 U4C
cloak *n.* 斗篷，外衣 U8A
cluster *n.* 串，丛，簇 U6A
coalesce *v.* 联合，合并 U9B
coarsely *adv.* 粗糙地 U7C
cognizant *a.* 认识到的 U1C
cohabitation *n.* 共生，共同存在 U6A
cohesive *a.* 有关联的 U2B
collaborate *v.* 协作，合作 U2C
collide *v.* 碰撞，冲突 U7C
combustible *a.* 可燃的，易燃的 U9C
commemorative *a.* 表示纪念的 U2A
commodity *n.* 便利，有用 U1C
complexity *n.* 复杂性 U2B
comprehend *v.* 理解，领会 U1A
concurrently *adv.* 并发地，同时地 U2A
conductance *n.* 传导 U10C
confound *v.* 使困惑，使不知所措 U7A
Confucianism *n.* 儒教，儒家学说或思想 U6C
conquest *n.* 征服 U8C
conscription *n.* 征兵，招兵 U9B
contention *n.* 论点 U5B
continuum *n.* 连续体 U7B
converge *v.* 会聚，汇合 U2B
coral *n.* 珊瑚 U8C
corrosive *a.* 腐蚀性的 U10C
cosmological *a.* 宇宙哲学的，宇宙论的 U6B
counteraction *n.* 消除，抵消 U7C
courthouse *n.* 县政府大楼 U7B
crane *n.* 鹤 U3C
crazed *a.* 疯狂的 U4A
critic *n.* 批评家，评论家 U2B

crude *a.* 不精细的 U7C

culminate *v.* 达到顶点 U7A

cultivated *a.* 人工栽培的 U2C

curtail *v.* 缩小 U9B

curvature *n.* 弯曲, 弯曲的形状 U10C

cypress *n.* 柏树 U6C

D

dazzling *a.* 令人赞叹不已的 U5A

dedicate *v.* 为（建筑）举行落成典礼 U1C

deem *v.* 认为 U6A

delicacy *n.* 优美, 精致 U5A

demonstrative *a.* 用于表明或说明的, 显示出的 U8B

denote *v.* 指示 U3C

depict *v.* 描绘; 描画 U7B

designate *v.* 清楚地标出或指出 U4A

dichotomy *n.* 一分为二（尤指成对立的两部分） U5B

diminutive *a.* 特小的 U9B

disciple *n.* 信徒, 门徒 U4B

disconcert *v.* 挫败, 使窘迫 U7A

disruption *n.* 分裂 U9C

diverging *a.* 分散的, 分开的 U3A

diverse *a.* 不同的 U6B

divine *a.* 神的 U3A

divinity *n.* 神, 神氏 U2C

doctorate *n.* 博士学位 U4B

domestic chores 家务杂活 U4A

domesticity *n.* 本土（特点） U5A

douse *v.* 在……上浇水 U3C

E

eclecticism *n.* 折中主义 U1A

eclosion *n.* 涌现 U6C

elaborate *a.* 复杂精美的 U2A

elasticity *n.* 弹性，活力 U10C

elevated *a.* 高贵的，抬高的 U1B

elliptical *a.* 椭圆形的 U3A

elusive *a.* 难以表述的 U1B

embodiment *n.* 具体表现，体现 U1B

embrace *v.* 包含 U6B

emulate *v.* 仿效 U3B

ephemeral *a.* 短暂的 U5B

epitaph *n.* 碑文，墓志铭 U4A

epitomize *v.* 成为……的典型范例 U2A

ethnical *a.* 种族的 U6A

ethos *n.* 社会（或民族等）的精神特质 U3B

evacuation *n.* 疏散，撤离 U9C

excavate *v.* 发掘（古物等） U2A

exceptionally *adv.* 非凡地，尤其地 U5B

expatriate *a.* 生活在国外的，被流放的 U3A

exquisite *a.* 精致的，精巧的 U6C

extoll *v.* 赞美，颂扬 U3B

exuberant *a.* 充满活力的 U2B

F

feudal *a.* 封建（制度）的 U6A

finalist *n.* 决赛选手或事物 U8C

flamboyant *a.* 艳丽的 U7C

flavour *n.* 特色 U6C

flint *n.* 燧石，火石 U10A

floral *a.* 花的，花图案的 U2B

flourish *v.* 繁荣，兴旺 U2A

foil *n.* 箔，金属薄片 U6A

foray *n.* 突袭 U4C

foreshadow *v.* 预示，预兆，预先给予暗示 U2B

formation *n.* 形状，结构 U2C

formidable *a.* 强大的，令人生畏的 U9B

formulate *v.* 规划，设计，构想 U2B

fortification *n.* 防御工事，要塞 U6B

fortress *n.* 堡垒，要塞 U7B

freehand *adv.* 不用绘图仪器地，徒手地 U1B

fuse *v.* 熔合，混合 U7A

G

garrison troop 卫戍部队，驻军 U6B

generic *a.* 普遍的，类属的 U5B

gigantic *a.* 巨大的，庞大的 U2A

glistening *a.* 闪耀的，光辉的 U8B

glitter *v.* 闪闪发光 U3C

graft *v.* 移植，嫁接 U4B

grandiose *a.* 宏伟的，壮丽的 U2A

gravesite *n.* 墓地，坟墓 U4A

grove *n.* 小树林，园林 U2C

guidepost *n.* 指示牌，路标 U5B

H

habitat *n.* 栖息地，居住地 U2C

hallmark *n.* 标志，特点 U6A

hath （古）= has U1C

heather *n.* 欧石南（杜鹃花科植物） U10A

Hellenistic *a.* 希腊风格的，希腊文化的 U8A

hew *v.* （hewn为过去分词） 砍，劈 U6C

hexagon *n.* 六角形，六边形 U6B

hierarchy *n.* 等级制度 U1B

hire on 接受雇用 U4A

honeycombed *a.* 蜂窝状的 U6C

honorary *a.* 荣誉的 U4B

horizontal *a.* 水平的 U3C

hovering *a.* 翱翔的，盘旋的 U4A

I

ideology *n.* 思想（体系） U1C
ignition *n.* 着火，点火 U9C
imperishable *a.* 不易腐坏的 U5B
inaugurate *v.* 开始，创始 U1A
incessant *a.* 不断的 U2C
indented *a.* 锯齿状的 U7B
indigenous *a.* 当地的，本土的 U2A
individuality *n.* 个性，独特性 U2B
infill *n.* 填充 U7C
infilling *n.* 填入，填入之物 U10B
ingeniously *adv.* 巧妙地 U3C
initiate *v.* 发起，开始 U2B
initiative *n.* 主动性，创造性 U1B
initiator *n.* 创始人 U4C
innovative *a.* 革新的，新颖的 U1B
installation *n.* 装置，设备 U3A
insulate *v.* 使绝缘 U10B
integral *a.* 构成整体所需要的 U2A
inter *v.* 埋葬 U4A
interpenetrating *a.* 相互渗透的 U4A
intersection *n.* 交汇，相交 U2A
intertwined *a.* 缠绕在一起的 U5A
intimacy *n.* 亲密性，隐私 U2B
intricate *a.* 错综复杂的，精心制作的 U5C
intriguing *a.* 引起兴趣的 U6B
intrinsic *a.* 内在的，固有的 U7A
invasion *n.* 入侵 U2C
Islam *n.* 伊斯兰教 U2B
isotropic *a.* 等向的 U10C

ivory *n.* 象牙 U8A

J

jail *v.* 关押，监禁 U8B

jasperware *n.* 碧玉细炻器 U3B

K

knight *v.* 封为爵士（或骑士） U4C

L

lace-like *a.* 像花边一样的 U5A

laminated *a.* 压层的 U10C

landmark *n.* 里程碑，界标 U7A

laureate *n.* 获奖者 U1C

lavish *a.* 奢侈的，过度的 U2C

legionnaire *n.* 军团士兵 U8B

literati *n.* 文人学士 U3C

liturgy *n.* 礼拜仪式 U3A

localization *n.* 地方化 U6C

longevity *n.* 长寿 U3C

longitudinal *a.* 纵向的 U10C

M

magnanimousness *n.* 宽宏大量，慷慨 U3C

majestic *a.* 庄严的，雄伟的 U5A

mammoth *a.* 巨大的 U3B

manifestation *n.* 显示，表明 U1A

marvel *n.* 奇迹 U7A

Mayan *a.* 马雅的，马雅人的，马雅文化的 U2A

medallion *n.* 大奖章，大勋章 U1C

meld *v.* 混合 U5A

mercantile *a.* 贸易的，经商的 U7B

millimeter *n.* 毫米 U5C

minimalism *n.* 极简抽象派艺术 U7A

moduli *n.* (modulus 的复数) 模具，系数 U10C

monastery *n.* 修道院 U2C

moorland *n.* 高沼地 U10A

mortuary *n.* 停尸房，太平间 U8C

mosque *n.* 清真寺 U2B

municipal *a.* 市政的，市的 U4B

mythical *a.* 神话的 U6A

N

naturalized *a.* 加入国籍的 U4B

Neolithic *a.* 新石器时代的 U6A

nodule *n.* 小圆块 U10A

nomad *n.* 游牧民 U8C

nomadic *a.* 游牧的，流浪的 U9A

nomination *n.* 提名，推荐 U1C

nominee *n.* 被提名者 U1C

nonrepresentational *a.* (美术) 抽象的 U1A

noteworthy *a.* 值得注意的 U2A

numeracy *n.* 计算能力 U1B

O

oblation *n.* 供品，祭品 U6C

obscure *a.* 不引人注意的 U5A

occult *a.* 超自然的，神秘的 U6A

octagon *n.* 八角形，八边形 U6B

offshoot *n.* 枝条，分枝 U9B

opulence *n.* 豪华，富裕 U3B

organic *a.* 有机的，有机物的 U2B

ornament *n.* 装饰品 U7C

ornamented *a.* 装饰的 U2B

ornate *a*. 装饰华丽的 U5C

oscillation *n*. 摆动, 动摇 U9C

outrageous *a*. 粗暴的 U7C

oval *a*. 椭圆形的 U2B

overlapping *a*. 重迭的, 交搭的 U4A

overt *a*. 公开的 U9B

P

palatial *a*. 宏伟的, 壮丽的 U6B

panorama *n*. 全景 U6B

papal *a*. 教皇的 U3A

paradox *n*. 自相矛盾的事物 U5B

partition *n*. 隔开物 U3C

patron *n*. 主顾, 业主, 资助人 U2B

patronage *n*. 资助, 赞助 U3A

perch *v*. 栖息, 坐落 U8C

perpendicular *a*. 成直角的, 垂直的 U10C

Peruvian *a*. 秘鲁的 U2A

petal *n*. 花瓣 U6B

pharos *n*. 灯塔 U8A

philosophically *adv*. 哲学地 U5B

phoenix *n*. 凤凰 U6A

Pope *n*. 教皇 U3A

porcelain *n*. 瓷器, 瓷 U3B

porous *a*. 多孔的, 渗水的 U8C

position *v*. 安放, 放置 U6A

precision *n*. 精确, 精确性 U2A

pre-Columbian *a*. 哥伦布到达美洲以前的 U2A

predominate *v*. 统治, 支配, 占优势 U2A

preeminent *a*. 卓越的 U3A

prelude *v*. 拉开序幕 U9C

prescriptive *a*. 规定的, 指示的 U4C

prestigious *a.* 有威望的，受尊敬的 U1C
prevail *v.* 流行，盛行 U2B
principle *n.* 法则，原则，原理 U2B
pristine *a.* 原始状态的，本来的 U3B
profligate *a.* 恣意挥霍的 U10A
proliferate *v.* 迅速大量地产生 U2A
proliferation *n.* 迅速增长，扩散 U2C
prolific *a.* 作品多的，多产的 U4A
prominence *n.* 显著，杰出 U2B
proprietor *n.* 所有人，业主 U1B
proto-modernist *a.* 典型现代主义的 U7B
prototype *n.* 原型 U2B
proviso *n.* （附带）条件，条款 U4B
punctuate *v.* 打断 U9B

Q

quadrangle *n.* 四边形 U3C

R

ram *v.* 撞击，夯入 U10A
rationalize *v.* 使合理化，合理地说明 U5C
ravage *v.* 破坏，蹂躏 U2C
rectangular *a.* 矩形的，具有矩形形状的 U2A
redeemer *n.* 救赎者，救世主 U8B
re-enforced *a.* 加固的 U8B
regal *a.* 宏伟的，豪华的，适合帝王的 U2B
reign *n.* 统治，（君主）统治时期 U2A
re-invigorate *v.* 再次激励 U4A
relic *n.* 舍利子，遗物 U6B
renovate *v.* 革新，刷新 U8C
repulse *v.* 击退，驱逐 U6A
resort to 采取，诉诸于 U6C

revere *v.* 尊敬，崇敬 U4C

revitalized *a.* 新生的 U2C

revival *n.* 复兴，重新流行 U1A

ruin *n.* 废墟，遗迹 U2B

rusticate *v.* 使成粗面石工 U7B

S

sacred *a.* 神圣的 U2C

scripture *n.* 经文 U6B

secular *a.* 世俗的 U6B

seep *v.* 渗漏 U9C

seismic *a.* 地震的 U9C

sepulcher *n.* 埋葬所，宗教圣物储藏所 U8C

serpentine *a.* 蜿蜒的 U2C

set off 衬托出 U2C

sheer *a.* 纯粹的 U3A

shimmering *a.* 闪闪发光的 U7A

shopping mall 商业街，商场，购物中心 U2C

shrubbery *n.* 灌木丛 U2C

slope *v.* 倾斜 U7C

sojourn *n.* 逗留 U10A

solicit *v.* 征集 U1C

spectrum *n.* 范围 U7A

spiderweb *n.* 蜘蛛网 U5C

spiked crown 带尖芒的王冠 U8A

splendor n. 壮丽，壮观，辉煌 U8A

spurt *v.* 喷射 U3C

stark *a.* 轮廓分明的，显眼的 U5A

stately *a.* 庄严的，宏伟的 U3B

stray *a.* 孤立的，零星的 U7B

streamlined *a.* 流线型的，现代型的 U7B

strife *n.* 冲突 U9B

stunning *a.* 令人吃惊的 U7A

surmount *v.* 覆盖在……顶上 U6B
sustaining *a.* 用以支撑的 U3A
symmetrical *a.* 对称的或呈匀称状的 U2A
symmetry *n.* 对称 U3A
synonymous *a.* 同义的 U7A
synthesis *n.* 综和，合成 U9A

T

tangential *a.* 正切的 U10C
Tantric *a.*（佛教）密教哲学的 U6B
Taoism *n.* 道教，道家学说 U6C
tapestry *n.* 织锦，挂毯 U3B
taunt *v.* 嘲笑，讥讽 U7A
tenaciously *adv.* 难以改变地 U1A
terminate *v.* 结尾 U3A
terminus *n.* 终点，边界 U9B
terrace *n.* 梯田 U2C
terrain *n.* 地形，地势 U2A
texture *n.* 手感，质感 U1A
thatch *n.* 茅草 U10A
theatricality *n.* 夸张，做作 U3A
threshold *n.* 起点，开端 U2B
throng *n.* 人群；大量；众多 U5C
time-honored *a.* 古老而受到尊重的 U3C
topographical *a.* 地形学的 U2C
traditionalist *n.* 传统主义者，因循守旧的人 U5C
transcend *v.* 超越 U5B
transitional *a.* 过渡时期的 U7B
travertine *n.* 石灰华 U7C
trinity *n.* 三位一体，三合一 U6A
triumphal *a.* 胜利的，凯旋的 U2A
trope *n.* 比喻 U7C
tumble *v.* 翻滚，流下 U8C

turf *n.* 草皮 U10A
turmoil *n.* 混乱 U4A

U

undulating *a.* 呈波浪形的 U3A
unexcelled *a.* 无可比拟的 U2A
unification *n.* 合一, 联合 U3A
unparalleled *a.* 无比的, 空前的 U2C
unprecedented *a.* 无前例的, 前所未闻的 U3A
unrefined *a.* 不精细的, 粗糙的 U7C
unsurpassed *a.* 无法超越的 U5B
upheaval *n.* 动乱, 剧变 U7A
utilitarian *a.* 实用的 U7A
Utopian *a.* 乌托邦的 U4C

V

vantage point (观察某物的) 有利位置 U1A
vegetation *n.* 植被 U6A
veritable *a.* 真正的, 名副其实的 U6B
versatility *n.* 多面性 U10C
viable *a.* 切实可行的 U9B
visualize *v.* 想象, 设想 U5B
vogue *n.* 流行, 时尚 U1A
volcanic *a.* 火山的 U8C

W

ward off 避开, 挡开 U6A
waterfront *n.* 滨水地区 U6B
wayfarer *n.* 旅客, 徒步旅行者 U6B
whirlwind *a.* 旋风般的, 快速的 U5A
worshipper *n.* 礼拜者, 崇拜者 U8B

Architectural Terms

A

acanthus *n.* 叶形装饰 U2A

adobe *n.* 砖坯 U10B

altar *n.* 祭坛，祭台 U4C

annexe *n.* 附属建筑 U6C

antechamber *n.* 前厅，接待室，香客室 U6C

aqueduct *n.* 高架渠，桥管输水道 U2A

arabesque *n.* 阿拉伯式花饰（由花果、蔓藤、几何图形等组成的精细图案，用作地毯花纹、壁画装饰等） U2B

arcade *n.* 拱廊 U2C

arch system 拱券体系 U1A

arena *n.* 竞技场（古罗马圆形露天竞技场的中心区域） U8B

art nouveau 新艺术（约1890–1910间流行于欧洲和美国的一种装饰艺术风格，以曲折有致的线条为其特色） U2B

articulation *n.* 接合，连接方法 U2A

asymmetry *n.* 不对称 U2B

atrium *n.* 天井，中庭 U1C

B

balustrade *n.* 栏杆，扶手 U3C

barrel vault 筒形（桶形）拱顶 U2A

basilica *n.* 长方形廊柱大厅式建筑 U1A

beam *n.* （建筑物等的）横梁 U3C

bearing wall 承重墙 U9C

belvedere *n.* 观景楼 U6B

bonding *n.* 接，接合 U6A

bracket *n.* 托架，支架，斗拱 U6A

brick-molded *a.* 砖模的 U9A

brutalism 粗野主义（一种展示未经装饰的巨大构件以表示结实和粗犷力量的建筑风格） U7C

building control officer 建筑管理官员 U1B

Byzantine *a.* 拜占庭式的 U1A

C

canopy *n.* 华盖，天篷 U5B
cantilever *v.* 如悬臂向外伸展 U5B
carving *n.* 雕刻 U5C
cast iron 铸铁 U2B
castellation *n.* 城堡形建筑 U7B
ceramics *n.* 制陶术，制陶业 U3B
château *n.* 法国式城堡、宫殿 U2C
circle template 圆形模板 U1B
civil engineering 土木工程(学) U4A
cladding *n.* 涂层，敷层 U7A
clapboard *n.* 墙板，隔板 U10A
cohesive type 粘合性 U1A
colonnade *n.* 列柱，柱廊 U3A
colossal *a.* (柱型)高大的 U3A
colosseum *n.* 古罗马椭圆形剧场 U2A
colossus *n.* 巨像 U8A
columnar building 带圆柱的建筑 U1A
complex *n.* 建筑群 U2C
Corinthian order 科林斯式柱型(有带叶形饰的钟状柱顶的柱型，为希腊柱型中最华丽者) U2A
cornice *n.* 檐口，飞檐 U7B
crucifix *n.* 十字架(型) U4A
cruck building 曲木屋架建筑 U10A
cubby *n.* 小房间 U6C
cupola *n.* 小穹顶 U8A
cylindrical *a.* 圆柱体的 U7C

D

diagonal *a.* 对角的，斜的 U5C
diagonal *n.* 斜构体，斜撑 U5C

dome *n.* 圆屋顶，穹顶 U1A

Doric order 多利斯柱型（纯朴、古老的希腊建筑风格） U2A

dormer window 天窗，老虎窗 U6B

E

eaves *n.* 屋檐，房檐 U3C

ecologically sustainable cities and communities 生态可持续型城市和社区 U1B

energy efficient building 节能建筑 U1B

entablature *n.* 古典柱式的顶部（由过梁、雕带和挑檐三部分组成） U2B

etching *n.* 蚀刻版画，蚀刻术 U3B

expressionist *a.* 表现主义的 U8C

F

facade *n.* （建筑物的）正面 U3A

facing *n.* （起保护作用的）面层，饰面 U10A

figurine *n.* 小雕像 U6A

flank *n.* 两侧，侧翼 U6C

flying buttress 飞券，飞拱 U2A

fort *n.* 堡垒，要塞 U9B

forum *n.* （古罗马城镇）用于公开讨论的广场（或市场） U2A

foyer *n.* 门厅 U5C

fresco *n.* 湿壁画 U3A

functionalism *n.* 功能主义建筑（主张把建筑的实用功能放在设计的首位） U4C

futurism 未来主义 U7C

G

gargoyle *n.* （哥特式建筑的）怪兽状的滴水嘴 U7B

glass-clad *a.* 玻璃覆盖的 U4B

glazed tile 釉面砖 U2B

glazed tiling 釉面砖，琉璃瓦 U6A

grid *n.* 棋盘式街道布局 U7B

groin vault 交叉拱顶，十字拱顶 U2A

grotesque *n.* (哥特式建筑的) 怪诞饰 U7B

grotto *n.* 岩洞, 洞穴 U6C

H

hipped *a.* 有斜脊的 U7B

homogeneous mass 均质体 U1A

horticulture *n.* 园艺 (学) U2C

I

ichnography *n.* 平面图 U5C

Ionic *a.* 爱奥尼亚 (式样) 的 (一种古老的希腊建筑风格) U2A

joist *n.* 托梁 U6A

K

keystone *n.* 拱顶石 U9A

kiln *n.* 窑 U10B

kiosk *n.* 凉亭, 小亭子 U6B

L

laid-out *a.* 设计的, 布局的 U8B

landscape architecture 景观建筑, 景观营造 U2C

lattice-work *n.* 格构, 格构制品 U8C

lead mortar 铅灰泥 U8A

ledge *n.* 壁架, 窗台 U5B

local authority planner 地方权威规划部门 U1B

locale *n.* 场所, 地点 U5B

M

mannerism 风格主义 (16世纪欧洲的一种艺术风格, 强调形式的奇巧, 风格上的个人癖好或对别人独特风格、技法的搬用和模仿; 亦称"矫饰主义"或"体裁主义") U2B

mansard roof 复折式屋顶, 折线型屋顶 U5C

marble *n.* 大理石 U2B

masonry *n.* 石工技艺，砖石建筑 U2A

mausoleum *n.* 陵墓 U5A

metallurgy *n.* 冶金学，冶金术 U5A

modernism 现代主义建筑风格 U2B

mortise-and-tenon 榫眼与榫舌 U3C

mosaic *n.* 马赛克图案，马赛克切割和镶嵌 U2A

motif *n.* 基本图案，基本色彩 U2B

multi-storey building 多层建筑 U9C

mural *n.* 壁画，壁饰 U6C

N

nave *n.* 中堂，正厅，正殿 U10A

neoclassicism 新古典主义（18世纪末、19世纪初流行的一种崇尚庄重典雅的建筑风格） U2B

O

obelisk *n.* 方尖碑 U3A

onion dome 洋葱头形屋顶（多见于俄罗斯东正教教堂） U8C

overhanging *a.* 突出的 U6A

P

pagoda *n.* 塔，宝塔 U6B

pane *n.* 长方块，长方格，窗格 U5C

pavilion *n.* 亭子，阁 U1C

pedestal *n.* 基座，底座 U8A

pediment *n.* 山花，三角墙（指希腊式古典建筑正面门廊顶上的装饰性三角形山头） U2B

pendentive *n.* 方墙四角圆穹顶支承拱 U1A

piazza *n.* （尤指意大利等城市中的）露天广场 U3A

pier *n.* 支墩 U1A

pillar *n.* 柱子 U3C

plaster *n.* 灰泥，灰浆 U4A

plasterwork *n.* 灰泥工艺 U10A

plinth *n.* 底座，柱基 U10A

pointed arch 尖券 U2A

pointed vault 尖拱 U2A

portico *n.* (有圆柱的) 门廊， 柱廊 U3B

post and lintel 柱子与横梁 U1A

postmodernism 后现代主义建筑风格 U2B

prairie house 草原住宅 U4A

prismatic *a.* 棱柱型的 U7C

R

rafter *n.* 椽 U6A

ramp *n.* 斜坡，坡道 U8A

rance *n.* 一种有蓝和白条纹的暗红色大理石 (产于比利时) U9A

real estate 房地产 U4B

recession *n.* 墙壁等的凹处，缩进处 U2B

reflective *a.* 反射的 U5C

relief *n.* 浮雕 U8A

residential *a.* 居住的，住所的 U4A

rib *n.* 拱肋 U1A

ribbed vault 肋拱 U2A

roof ridge 屋脊 U3C

roofline *n.* 屋顶轮廓 U7C

rotunda *n.* 圆形建筑， 圆形大厅 U3B

rubble *n.* 碎石 U10B

Russo-Byzantine 俄式拜占庭风格的 U8C

S

sculptural *a.* 雕刻的，雕塑的 U4C

setback *n.* (墙壁上部厚度减低形成的) 壁阶，缩进 U7B

shaft *n.* 柱身 U2A

sheathe *v.* 覆盖，套装 U1A

shrine *n.* 神殿，神龛，圣祠 U2A

silica *n.* 硅石，硅土 U10A

skyscraper *n.* 摩天大楼 U1C

slab *n.* 平板，平板状物 U5B

slate *n.* 石板，石片瓦 U10A

span *v.* 横跨 U8A

spiral volute 螺旋涡形，涡形花样 U2A

spire *n.* 尖顶，塔尖 U6B

spirit gate 神门 U6A

steel-skeleton system 钢骨体系 U1A

stele *n.* 石碑 U6B

structural expressionism 结构表现主义 U7C

stucco *n.* （涂建筑物外墙用的）灰泥 U2A

stupa *n.* （印度）佛塔，浮屠塔 U6B

substructure *n.* 基础，下层结构 U1A

subterranean *a.* 地下的 U5C

sun-baked earth 日光烘培土 U9A

surveyor *n.* 测量员，检查员 U1B

T

tablet *n.* 牌匾，匾额 U6B

tapering *a.* 锥形、锥体的 U8A

thrust type 嵌入式 U1A

tier building 多层建筑 U6A

tile *n.* 瓷砖 U3C

tilework *n.* 砖瓦工艺 U2C

timber *n.* 木材，木料 U3C

toukong 斗拱（中国建筑特有的一种结构。在立柱和横梁交接处，从柱顶上加的一层层探出成弓形的承重结构叫拱，拱与拱之间垫的方形木块叫斗，合称斗拱） U3C

town hall 市政厅 U7B

trabeated system 横梁式结构 U1A

transparency *n.* 透明性，透明度 U5C

triumphal arch （为纪念胜利而建的）凯旋门 U2A

truss work（支撑屋顶，桥等的）构架 U7C

turquoise pencil 绿松石铅笔 U1B

turret *n.* 角楼，塔楼 U7B

Tuscan *a.* 托斯卡纳柱型的（古罗马建筑中的一种柱子的式样） U3A

U

unadorned *a.* 未装饰的，朴实的 U3B

upturned eaves 尖端向上翻的檐，挑檐 U3C

V

vaulted roof 拱形屋顶 U1A

vaulting *n.* 拱顶营造，造拱术 U2A

ventilation *n.* 通风，空气的流通 U5C

veranda *n.* 游廊，走廊，阳台 U6B

void *n.* 孔隙 U1A

voussoir *n.* 楔形拱石 U9A

W

watch tower 瞭望塔（烽火台） U9B

wooden truss 木质构架 U1A

Proper Names

A

Abu Simbel 阿布辛拜尔城（埃及） U9A

Adler and Sullivan 阿德勒和沙利文（建筑事务所） U4A

Aegean Sea 爱琴海 U8A

Agra 阿格拉（印度北部城市，泰姬陵所在地） U2B

Agrigento 阿格里琴托（意大利西西里岛西南的一城市，俯瞰地中海） U2A

AIA (the American Institute of Architects) 美国建筑师协会 U4B

Alexandria 亚历山大港(埃及北部港市) U8A

Alhambra 爱尔罕布拉宫（古迹，在西班牙南部城市格拉纳达，是13–14世纪时的摩尔人宫殿） U2B

Amazon 亚马逊族(相传曾居住在黑海边强壮的女战士一族) U8A

Amiens 亚眠(法国北部城市) U1A

Andes 安第斯山脉(南美洲西部巨大山脉) U2A

Andes Plateau 安第斯高原(南美洲) U8B

Andrea Palladio 安德烈亚·帕拉迪诺(1508–1570, 意大利建筑师,常被认为是西方最具影响力的建筑师) U5A

Angkor Wat 吴哥窟(吴哥为柬埔寨西北的重要考古遗址) U8C

Anthony van Dyck 安东尼·凡·戴克(1599–1641, 比利时画家) U3A

Antoni Gaudí 安东尼·高迪(1852–1926, 西班牙著名建筑师,被尊为有机主义建筑流派开创人) U2B

Antonio Sant'Elia 安东尼奥·桑塔埃利亚(意大利著名建筑设计师) U7C

Aqueduct 高架输水桥(古罗马) U9A

Arch of Titus in Rome 提图斯凯旋门 U9A

Art Deco 装饰派艺术(起源于20世纪20年代的装饰和建筑艺术风格,以轮廓和色彩明朗粗犷、呈流线型和几何形为特点) U7A

Artemis 阿耳忒弥斯女神(希腊神话中狩猎女神和月神) U8A

Arts and Crafts Movement 艺术与工艺运动(19世纪后期至20世纪早期兴起于英美的工艺美术运动) U4A

Asia Minor 小亚细亚(黑海与地中海之间亚洲西部的一个半岛,总体上与土耳其范围相当) U2A

Assyrian 亚述人(古代生活在两河流域上游的民族,建立了亚述帝国) U1A

AT&T Building 美国电话电报公司大厦(纽约) U2B

Athens 雅典(希腊的首都) U2A

Augustus 奥古斯都(63 BC– AD 14, 罗马帝国第一代皇帝) U1C

Australian Aborigines 澳洲土著居民 U9A

Austrian Benedictine Abbey 奥地利本笃会修道院(本笃会为基督教分支教派团体) U2B

B

Bagua style 八卦式 U6C

Baldassare Peruzzi 巴尔达萨雷·佩鲁齐(1481–1536, 意大利文艺复兴盛期建筑师和画家,1520年被任命为罗马圣彼得大教堂的建筑师) U2B

Balthasar Neumann 约翰·巴塔萨·纽曼(1687–1753, 德国建筑师) U3A

Baltic coast 波罗的海沿岸 U10B

Bank of China 中国银行 U4B

Barcelona 巴塞罗那（西班牙东北部港市） U2B

Baroque style 巴洛克式风格（约1550到1700年间盛行于欧洲的一种建筑风格，强调拉紧的效果，其特征是有粗的曲线结构、复杂的装饰和无联系部分间的整体平衡） U1A

Bauhaus (school) 包豪斯建筑学派（德国建筑研究学派，或指其风格） U1A

Bavaria 巴伐利亚（州）（德国南部地区，原为公爵领地） U2B

Bombay 孟买（印度西部港市） U2A

Brighton 布赖顿（英国英格兰东南部城市） U3B

Bronze Age 青铜器时代 U9A

Brussels 布鲁塞尔（比利时首都和最大城市） U2B

Buddhist architecture 佛教建筑 U6C

Burgundy 勃艮第（法国中东部地区） U2A

C

Callicrates 卡立克拉特（希腊建筑师，雅典卫城的主要设计师） U8C

Calvert Vaux 卡尔弗特·沃克斯（1824–1895，英裔美国景观设计师，纽约城内的中心公园的设计者之一） U2C

Cambodia 柬埔寨（东南亚国家） U8C

Cangjingge 藏经阁 U6B

Canterbury 坎特伯雷（英格兰东南部城市） U2A

Canton 广州 U4B

Caravaggio 米开朗琪罗·卡拉瓦乔 U3A

Caria 卡里亚（小亚细亚西南部濒临爱琴海的一古老地区，曾为波斯帝国行省） U8A

Carlo Maderno 卡罗·玛丹诺（意大利建筑师，以巴洛克风格著称） U3A

Carracci 卡拉齐画派（他们的作品影响并导致巴洛克式艺术过渡期的风格主义的变革） U3A

Carrousel du Louvre 罗浮宫旋转木马 U3B

Casa Milά 米拉公寓 U2B

Cass Gilbert 凯斯·吉尔伯特 U7A

Castle of the White Heron 白鹭城（日本姬路） U9A

Cathedral Square 教堂广场（在克里姆林宫内） U8C

Centaur （希腊神话）人首马身怪 U8A

Chaldaean 迦勒底人（与巴比伦人血缘相近的闪米特人） U1A

Chambord 商堡（法国中北部一村镇，以文艺复兴时期法王弗朗西斯一世在此兴建的壮观华美的城堡而

闻名) U2C

Champs-Élysées 爱丽舍宫 (巴黎古建筑, 法国总统官邸, 建于18世纪初) U3B

Charles Bulfinch 查尔斯·布尔芬奇 U3B

Chartres Cathedrals 夏特尔大教堂 (法国) U1A

Château de Versailles (法国) 凡尔赛宫 (欧洲最大王宫, 凡尔赛为法国中北部一个城市, 位于巴黎西南以西) U2B

Chersiphron 切西弗龙 (古希腊著名建筑师, 建造了阿耳忒弥斯神庙) U8A

Chicago school 芝加哥学派 (美国建筑研究学派, 或指其风格) U2B

Chicago Tribune Tower 芝加哥论坛报大厦 U7A

Chili 智利 (南美洲西南部国家) U8C

Christian theology 基督教义, 基督神学 U2A

Church of Hagia Sophia 圣索菲娅大教堂 (在土耳其伊斯坦布尔市, 原为拜占庭帝国东正教的宫廷教堂, 拜占庭建筑风格的代表作) U1A

Church of St. Andrew 圣安德鲁大教堂 (英国埃塞克斯郡) U10A

Church of the Holy Family 神圣家族大教堂 U2B

Claude Lorrain 克劳德·洛兰 (1600–1682, 法国风景画家, 开创表现大自然诗情画意新风格) U3A

Claude Nicholas Ledoux 克劳德·尼古拉斯·雷多 U3B

Compton Wynyates, Warwickshire 康普顿宅邸 (英国沃里克郡) U10B

Comtesse du Barry 巴里伯爵夫人 U3B

Confucius Temple 孔庙 U6B

Constantine the Great 康斯坦丁大帝 (古罗马皇帝, 288–337在位, 建立了君士坦丁堡) U2A

Constantinople 君士坦丁堡 (拜占庭帝国首都, 现为土耳其西北部港市伊斯坦布尔) U1A

Corcovado Mountain 科尔瓦多山 (毗邻里约热内卢市) U8B

Córdoba 科尔多瓦 (西班牙城市, 曾经为阿拉伯倭马亚王朝都城) U2B

Croesus 克利萨斯 (吕底亚王国的末代国王, 以富有著称) U8A

Crystal Palace (伦敦) 水晶宫 (第一届世界博览会会址) U2B

D

Da Cortona 达·科托纳 (意大利建筑师、画家, 巴洛克艺术的倡导者) U3A

Dandong 丹东 U9B

Delhi 德里 (印度中部偏北城市, 自古以来该市就占重要地位, 17世纪沙·贾汗皇帝重建这座古城围住莫卧儿皇宫) U2B

Domenico Ghirlandaio 多米尼哥·吉兰达约 (1449–1494, 文艺复兴初期弗洛伦萨画家, 曾为梵蒂冈西斯廷礼拜堂作画) U4C

Domitian 图密善(古罗马皇帝, 81–96年在位) U9B

Domus Augustana 奥古斯塔纳圆顶殿 (意大利) U9B

Dorset 多塞特郡 (英格兰南部) U10A

Durham Cathedral 都汉姆大教堂 (英国) U2A

E

Easter Island 复活节岛 (智利) U8C

École des Beaux-Arts (兴起于法国的) 装饰艺术风格派 U1A

Edinburgh 爱丁堡 (英国苏格兰首府) U3B

Efanggong 阿房宫 U6B

Eiffel Tower (巴黎) 埃菲尔铁塔 (在塞纳河南岸) U2B

Elgin Marbles 埃尔金大理石雕 (指一些雅典雕刻及建筑残件, 于19世纪由英国伯爵Thomas Elgin运至英国, 现藏不列颠博物馆) U3B

Ellora & Ajanta 埃罗喇和阿旃陀 (位于印度中西部的古村, 附近岩洞可追溯到约公元前200年至公元650, 藏有辉煌的佛教艺术典范) U2A

Ephesus 以弗所 (小亚细亚西岸的贸易城市, 以阿耳忒弥斯神庙而闻名) U8A

Essex 埃塞克斯郡 (英格兰东北部) U10A

Etruscan (意大利中西部古国) 伊特鲁里亚的, 伊特鲁里亚人的 U3B

Exeter 埃克塞特 (英格兰西南部城市) U2A

F

Fallingwater 流水别墅 U4A

Federal style 美国联邦时期风格 U2B

Filippo Brunelleschi 菲利波·布鲁内莱斯基 (1377–1446, 意大利文艺复兴初期建筑师, 代表作有圣洛伦佐教堂和弗洛伦萨的圣马利亚教堂) U2B

Flaxman 弗拉克斯曼 (1755–1826, 英国雕刻家) U3B

Florence Cathedral 佛罗伦萨大教堂 U2B

Florentine 佛罗伦萨 (人) 的; 佛罗伦萨文化的 U4C

Fragrant Hill Hotel (北京) 香山饭店 U4B

Francesco Borromini 弗朗切斯科·博罗米尼 (1599–1677, 意大利建筑家) U3A

Frank Gehry 弗兰克·葛瑞(普利兹克建筑奖得主) U1C

Frank Lloyd Wright 弗兰克·劳埃德·赖特(1869–1959, 美国著名建筑师, 他基于自然形式的特殊建筑风格极大地影响了现代建筑业) U1C

Frederic Auguste Bartholdi 弗雷德里克·奥古斯特·巴托尔蒂(法国著名雕刻家, 自由女神像的创造者) U8C

Frederick Law Olmsted 弗雷德里克·劳·奥姆斯特德(1822–1903, 美国景观设计师, 纽约市中央公园的主设计师) U2C

G

Georgian style 乔治王朝时期建筑风格 U1A

Giacomo da Vignola 贾科莫·达·维尼奥拉(1507–1573, 意大利建筑大师, 风格主义建筑代表之一, 其罗马耶稣会教堂的十字形设计对后世的教堂建筑有很大的影响, 著有《建筑五柱式规范》等) U2B

Gianlorenzo Bernini 乔洛伦佐·伯尼尼(1598–1680, 意大利雕塑家、画家和建筑家, 意大利巴洛克风格的杰出代表, 以其流畅、动感的雕塑及其为包括圣彼得大教堂在内的许多教堂的设计而著称) U2B

Giovanni Battista Piranesi 乔万尼·巴迪斯塔·皮兰尼西 U3B

Giulio Romano 朱利奥·罗马诺(1499? –1546, 意大利文艺复兴晚期画家和建筑家, 拉斐尔的学生, 意大利风格主义奠基人, 曾设计圣贝内德托教堂) U2B

Giza 吉萨(埃及东北部城市, 同开罗隔河相望; 南郊8公里之利比亚沙漠中有著名的金字塔、狮身人面像等古迹) U2A

Gothic architecture 哥特式建筑(流行于西欧的12世纪到15世纪的一种建筑风格, 特征是有尖角的拱门, 肋形拱顶和飞拱) U1A

Granada 格拉纳达(西班牙南部城市, 位于科尔多瓦东南, 由摩尔人于8世纪创建) U2B

Great Mosque at Djenne 德贾尼大清真寺(西非) U9A

Greek revival (19世纪上半叶盛行的)希腊复兴式 U2B

Guarino Guarini 加里诺·加里尼(1624–1683, 意大利建筑家) U3A

Gustave Eiffel 古斯塔夫·埃菲尔(1832–1923, 法国工程师, 为1889年的巴黎世界博览会设计埃菲尔铁塔) U8C

H

Hadrian 哈德良(古罗马皇帝, 117–138年在位) U2A

Hadrian's Villa at Tivoli 古罗马哈德良皇帝宅邸(意大利) U9B

Halicarnassus 哈利卡那苏斯(位于小亚细亚今天土耳其境内西南部爱琴海海滨的希腊文化古城) U8A

Hampton Court Palace 汉普敦王宫(英国) U10B

Han Dynasty 汉朝 U6A

Hanseatic Towns (中世纪)商业同业公会的市镇 U10B

Hearst Castle 赫斯特堡(威廉·伦道夫·赫斯特在美国加利福尼亚圣西蒙为自己修建的富丽堂皇的庄园) U9B

Heitor da Silva Costa 黑托·达·谢尔瓦·科斯塔(巴西设计师) U8B

Helios (希腊神话)太阳神赫利俄斯(相当于罗马神话的阿波罗) U8A

Hengrave Hall, Suffolk 亨格瑞夫宅邸(英国萨福克郡) U10B

Henry Moore 亨利·穆尔(1898–1986,英国雕刻家,按照自然形体和节奏原则而非几何形体作抽象雕刻,代表作有石雕《母与子》等) U1C

Henry Wotton 亨利·沃顿(1568–1639,英国外交家及诗人) U1C

Henry-Russel Hitchcock 亨利·拉塞尔·希契科克(建筑史学家) U5B

Herculaneum 海格力古城(意大利南部古城) U3B

Herodotus 希罗多德(古希腊著名历史学家) U8A

Holy Trinity Church in Hull 赫尔市圣三一教堂(英国) U10B

Howells 豪威尔斯 U7A

Humphry Repton 亨弗利·雷普顿(18世纪英国著名景观设计师) U2C

Hungary 匈牙利(欧洲中部国家) U1A

Hyatt Regency Atlanta 海厄特行政大厦(亚特兰大市) U1C

I

Ictinus 伊克蒂诺(希腊建筑师,雅典卫城的主要设计师) U8C

Ieoh Ming Pei 贝聿铭 U4B

Imperial Hotel (东京)帝国饭店 U4A

Inca 印加人(秘鲁高原上克丘亚部落的一支,被西班牙征服前建立了一个北起厄瓜多尔南到智利中部的帝国) U2A

Industrial Revolution 工业革命(发生于18世纪末期英国) U2B

International style 国际风格 U1A

Ivan IV 伊凡四世(伊凡雷帝,俄罗斯沙皇,带领俄罗斯人摆脱鞑靼人的统治,以暴烈著称) U8C

J

James Stuart 詹姆士·斯图亚特 U3B

Jay A. Pritzker 杰伊·普利兹克(芝加哥富商,普利兹克奖的创办者) U1C

Jean-François Chalgrin 让·弗朗索瓦·查尔金 U3B

Jianguomen 建国门(北京) U6B

Jimmy Carter 吉米·卡特(美国第39任总统) U4B

Jin Dynasty 晋朝 U6C

Johann Bernhard Fischer von Erlach 约翰·伯恩哈德·菲歇尔·冯·厄拉策(1656–1723, 奥地利建筑师) U3A

Johann Winckelmann 约翰·温科尔曼 U3B

John F. Kennedy Library 肯尼迪总统图书馆 U4B

John Hancock Center 约翰·汉考克大厦(美国芝加哥) U7C

John Nash 约翰·纳什 U3B

Johnson Administration Building 约翰逊(制蜡公司)办公大楼 U4A

Jørn Utzon 约翰·伍钟(20世纪丹麦最重要的建筑设计师, 其最著名的设计是悉尼歌剧院) U8C

Justinian I 查士丁尼一世(483–565, 东罗马帝国皇帝, 527–565在位) U2A

K

Karl Friedrich Schinkel 卡尔·弗莱德里西·辛克尔 U3B

Katsura Detached Palace 桂离宫(日本) U2A

Kazan 喀山(俄罗斯欧洲部分的东部城市, 位于莫斯科东部, 乌拉尔河上) U8C

Khafre 哈夫拉(古埃及第四王朝国王, 于公元前约2530年建哈夫拉金字塔和狮身人面像) U2A

Khmer 高棉人(柬埔寨的一个民族, 其文明在9世纪至15世纪时达到高峰) U8C

Khorsabad 霍萨巴德(古亚述帝国的一个城市) U2A

Khufu 胡夫(古埃及第四王朝第二代国王, 下令建造最大的金字塔) U2A

Kiyomizu Temple 清水寺(日本京都) U8C

Krak des Chevaliers in Syria 十字军城堡(叙利亚) U9A

Kremlin 克里姆林宫(历代俄罗斯沙皇的宫邸, 现为俄国政府办公所在地) U8C

Kuiwenge 魁文阁 U6B

Kyoto 京都(日本古都) U8C

L

Lake Shore Drive Apartments 滨湖大道公寓楼(雷克·肖大道公寓楼, 在芝加哥) U2B

Lapith 拉毗士(古希腊塞萨利地区的部落) U8A

Lascaux 拉斯考克斯(法国西南部一山洞) U9A

Le Corbusier 勒·科比西埃(1887–1965, 瑞士裔法国建筑师, 重要作品有马赛公寓等) U2B

Liao Dynasty 辽代(朝) U6C

Lincoln 林肯市(英格兰东部城市) U2A

Lisbon 里斯本(葡萄牙首都和最大的城市) U8B

Little Wenham Hall in Suffolk 萨福克郡小温海姆宅邸(英国) U10B

Loire Valley 卢瓦尔河谷(卢瓦尔河是法国最长的河流, 位于法国中部) U2C

Lombard Romanesque 伦巴第罗马式 U1A

Longmen Grottoes 龙门石窟 U6C

Louis H. Sullivan 路易斯·苏利文(1856–1924, 美国建筑师, 弗兰克·赖特之师, 芝加哥学派的代表人物之一, 主张"功能决定形式", 主要作品有芝加哥的会堂大厦、圣路易斯的10层温赖特大厦等) U1C

Louis XIV 路易十四(1638–1715, 法国国王, 1643–1715年在位, 绰号太阳王) U2B

Louveciennes 卢维谢纳 (法国城市) U3B

Ludwig II 路德维希二世(德国巴伐利亚的末代君主, 修建了新天鹅城堡) U8C

Lydia 吕底亚(小亚细亚中西部一古国, 以其富有和奢华而闻名) U8A

Lyndon Johnson 林登·约翰逊(美国第36任总统) U4B

M

Machu Picchu 马丘比丘古城(遗迹, 坐落在秘鲁境内安第斯高原上) U8B

Mali 马里(西非国家) U8C

Marcus Vitruvius 马库斯·维特鲁威(公元前1世纪, 古罗马建筑师, 所著《建筑十书》在文艺复兴时期、巴洛克及新古典主义时期成为古典建筑的经典) U1C

Marduk 马杜克(古代巴比伦人的主神, 原为巴比伦的太阳神) U8A

Marin County Civic Center 马林文娱中心 U4A

Massachusetts State House 马萨诸塞州议会大厦 U3B

Mausolus 摩索拉斯王(波斯帝国卡里亚省的总督, 其陵墓非常壮观) U8A

Maya 马雅人(中美洲古印第安人的一族, 具有高度文明) U2A

McGraw-Hill Building 麦克格劳–希尔大厦 U7A

Medici Villa 美第奇家宅(美第奇是意大利文艺复兴时期统治佛罗伦萨的贵族, 曾赞助过米开朗琪罗等艺术家) U2C

Mediterranean 地中海 U8A

Mesopotamia 美索不达米亚(古代西南亚介于底格里斯河和幼发拉底河之间的一个地区, 位于现在的伊拉克境内) U2A

Michelangelo 米开朗琪罗(1475–1564，意大利文艺复兴盛期成就卓著的雕刻家、画家、建筑师，建筑设计罗马圣彼得大教堂圆顶) U2B

Mies van der Rohe 密斯·凡·德·洛(著名德裔美国建筑师) U1C

Miho Museum (日本)美秀博物馆 U4B

Millennium Park 千禧公园(芝加哥) U1C

MIT 麻省理工学院 U4B

Mitterrand 密特朗(法国前总统) U5C

Moai 摩艾(复活节岛巨像的别称) U8C

Mogao Crottoes 莫高窟 U6C

Mongolian Plain 蒙古草原 U8B

Monticello 蒙蒂塞洛(美国弗吉尼亚州中部夏洛茨维尔东南一住宅区，由托马斯·杰斐逊设计) U3B

Moor 摩尔人(非洲西北部阿拉伯人与柏柏尔人的混血后代，于公元8世纪进入并统治西班牙) U2C

N

Nabataean Empire 纳比第安帝国 U8B

National Center for Atmospheric Research (美国)国家大气研究中心 U4B

National Gallery of Art (美国)国家美术馆 U4B

Nebuchadnezzar II 尼布甲尼撒二世(古巴比伦国王，攻占耶路撒冷，建空中花园) U8A

Neuschwanstein Castle 新天鹅城堡(坐落在德国境内巴伐利亚省的施旺高) U8C

Nicholas Revett 尼古拉斯·利维特 U3B

Nike Apteros (希腊神话)胜利女神 U2A

Nîmes 尼斯 (法国南部城市) U3B

Ningbo 宁波 (浙江) U6B

Norham Hall Essex 诺汉姆宫(英国埃塞克斯郡) U10B

Norman 诺曼人(10世纪定居于诺曼底的斯堪的纳维亚人和法国人的后裔) U10A

Northern Wei Dynasty 北魏朝 U6C

Notre Dame (法语)圣母 U2A

O

Olympia 奥林匹亚(希腊南部平原，用以祭拜宙斯的宗教中心，古代奥林匹克运动会的遗址) U8A

Osterley Park 奥斯特雷公园 U3B

Ottobeuren 奥托伯伦堡(奥地利城市) U2B

P

Palace at Knossos 克诺索斯宫殿 (希腊克里特岛) U9A

Palazzo Vecchio 韦基奥宫 (意大利佛罗伦萨) U7B

Palladio's Basilica 帕拉迪诺教堂 U5A

Parthenon 帕台侬神殿 (希腊神话中女神雅典娜的主要神庙，位于雅典卫城上，建于公元前447年和公元前432年之间，多利安式建筑的杰出代表) U2A

Paul Landowski 保罗·朗多斯基 (法国雕塑家) U8B

Pericles 伯里克利 (古雅典领袖，因其推进了雅典民主制并下令建造帕台侬神庙而著名) U8C

Persian 波斯人 (古代或中世纪) U2C

Petra 佩特拉古城 (约旦) U8B

Pharos 法罗斯岛(位于埃及亚历山大港的地中海上，后来指代灯塔) U8A

Phidias 菲迪亚斯 (古希腊雅典雕塑家，奥林匹亚的宙斯雕像是其杰作) U8A

Philadelphia 费城 (美国宾西法尼亚州东南部港市) U3B

Pierre Lescot 皮埃尔·莱斯科 (法国建筑师) U5A

Polynesian 玻利尼西亚人 (中太平洋岛屿原始居民) U8C

Pompeii 庞培古城 (意大利南部古城，公元79年火山爆发，全城淹没) U3B

Pons Fabricius 伐布里修斯桥 (古罗马) U9A

Pont du Gard 伽合大桥 (法国) U9A

Poseidon (希腊神话) 海神波塞冬 U8A

Poussin 尼古拉斯·普珊 (1594–1665, 法国画家) U3A

Prairie school 草原学派 (19世纪末20世纪初以赖特为首的芝加哥年轻建筑师共同打造的学派，探究建筑的水平发展) U4A

Price Tower 普赖斯大厦 U4A

Protestant 新教徒 U3A

Q

Qin Dynasty 秦朝 U6B

Qufu 曲阜 (山东) U6B

R

Raffles City 莱福士广场 U4B

Ramesses II 拉美西斯二世(古埃及法老) U9A

Raymond Hood 雷蒙德·胡德 U7A

Regency style (英国)摄政时期风格 U3B

Regent Street 摄政大道(伦敦) U3B

Reims 兰斯(法国东北部城市) U2A

Rembrandt 伦勃朗(1609–1669, 荷兰著名画家,以肖像画著名) U3A

Renaissance architecture 文艺复兴时期风格的建筑 U1A

Residenz at Wuzburg 坐落在伍兹伯格的宅邸 U9B

Rhodes 罗得岛(位于爱琴海畔东南部) U8A

Richmond 里士满 (美国弗吉尼亚州首府) U3B

Rio de Janeiro 里约热内卢市(巴西东南部港市) U8B

Robert Adam 罗伯特·亚当 U3B

Rococo style 洛可可式(18世纪初起源于法国、18世纪后半期盛行于欧洲的一种建筑装饰艺术风格,其特点是精巧、繁琐、华丽) U2B

Romanesque architecture 罗马风建筑(包含古罗马和拜占庭特色的欧洲建筑风格,该风格盛行于11世纪和12世纪,特点为包括厚实的墙、筒拱穹顶及相对不精细的装饰品) U1A

Royal Pavilion at Brighton 布赖顿皇家观景阁 U3B

Rubens 彼得·保罗·鲁本斯(1577–1640, 比利时画家,巴洛克艺术代表) U3A

Rumania 罗马尼亚 U10A

Russian Orthodox 俄罗斯东正教 U3A

S

Saint Peter's Basilica 圣彼得大教堂(教皇庭,天主教的中心,坐落在梵蒂冈城) U2B

Salisbury Plain 索尔兹伯里平原 U8C

Sāmarrā 撒马拉(伊拉克中北部城市,位于巴格达西北偏北部的底格里斯河畔) U2B

Sant' Agnese Fuori le Mura 圣阿格涅斯教堂 U2A

Santa Sabina 圣萨比纳教堂 U2A

Santiago de Compostela in Spain 圣地亚哥大教堂(西班牙) U9A

Sargon II 萨尔贡二世(古亚述帝国的君主) U2A

Schauhaus (德国历史博物馆的)展馆;展室 U4B

Schwangau 施旺高（德国巴伐利亚省南部一地区，临近图森镇） U8C

Seagram Building 西格拉姆大厦（纽约） U2B

Sèvres 塞夫勒（法国北部城市） U3B

Shakespeare's Globe Theatre 莎翁环球大剧院 U9A

Shang Dynasty 商朝 U6A

Shanhaiguan 山海关 U9B

Sicily 西西里（意大利南部岛屿，位于意大利半岛南端以西的地中海） U2A

Siem Reap 塞姆瑞菩村（柬埔寨） U8C

Sir Christopher Wren 克里斯托弗·雷恩爵士（1632–1723，英国著名建筑师，1666年伦敦大火后设计圣保罗教堂等50多座伦敦教堂，还有许多宫廷建筑、图书馆等） U2B

Sir John Soane 约翰·索恩爵士 U3B

Sir Joseph Paxton 约瑟夫·帕克斯顿爵士（英国著名建筑师） U2B

Sir Robert Smirke 罗伯特·斯莫克爵士 U3B

Sistine Chapel 西斯廷礼拜堂（梵蒂冈的主要教堂，以米开朗琪罗及其他艺术家的天顶画和壁画著称） U4C

Song Dynasty 宋朝 U6C

Southern and Northern Dynasties 南北朝 U6C

St. Basil's Cathedral 圣巴兹尔大教堂（坐落在莫斯科红场，为俄拜占庭式建筑著名代表作） U8C

St. Denis Church 圣丹尼斯教堂 U5A

St. Patrick's Cathedral 圣巴特里克大教堂 U1A

St. Susanna 圣苏珊纳大教堂 U3A

Statue of Liberty 自由女神像（纽约） U8A

Stonehenge 巨石阵（英国南部索尔兹伯里附近的一组立着的石群） U8C

Sui Dynasty 隋朝 U6C

Suryavarman II 苏利阿伐曼二世（12世纪上叶高棉帝国的国王） U8C

Sutra Library 藏经室 U6C

Syon House 西昂之宅 U3B

T

Taj Mahal 泰姬陵（位于印度阿格拉市，国王沙·贾汗在1629年为其妃所建的陵墓） U1A

Taliesin Fellowship 塔里埃森设计团体 U4A

Taliesin West 西塔里埃森 U4A

Taliesin 塔里埃森 U4A

Tang Dynasty 唐朝 U6A

Taoist architecture 道教建筑 U6C

Tatar 鞑靼人 U8C

Teotihuacán 提奥帝华坎（位于今天墨西哥城东北的中墨西哥的古城，其遗址包括太阳金字塔和克萨尔科多神庙） U2A

the Acropolis 雅典卫城 U2A

the British Museum 大英博物馆 U3B

the Chicago Fire 芝加哥大火 U7B

the Four Seasons Hotel 四季酒店 U4B

the German Historical Museum 德国历史博物馆 U4B

the Grand Louvre Pyramids （法国）卢浮宫金字塔 U4B

the Guggenheim Museum 古根海姆美术馆（纽约） U4A

the Hyatt Foundation 海厄特基金会 U1C

the Maison-Carrée 伽黑之宅 U3B

the Morton H. Meyerson Symphony Center 梅尔森音乐厅 U4B

the Pritzker Architecture Prize 普利兹克建筑奖 U1C

the Royal Theater in Berlin 柏林皇家剧院 U3B

the Silk Road 丝绸之路 U6B

the State Capitol Building 美国国会大厦 U3B

the Wainwright Building 韦恩赖特大厦 U5A

the Yalu River 鸭绿江 U9B

Theodorus 西奥多罗斯（古希腊建筑师，重建了阿耳忒弥斯神庙） U8A

Thomas Jefferson 托马斯·杰斐逊（1743–1826，美国第三任总统） U3B

Tianyige 天一阁 U6B

Tiepolo 乔瓦尼·巴蒂斯特·泰波罗（1696–1770，意大利画家） U3A

Timbuktu 廷巴克图（马里中部城市） U8C

Titus 提图斯（古罗马皇帝，统治期间继续修建罗马圆形大竞技场） U8B

Tivoli 提沃利（意大利中部城市，位于罗马东北偏东） U2A

Toledo 托莱多（西班牙中部城市，位于马德里西南偏南，曾经是摩尔人的首都） U2C

Tomb of Humayun 胡马雍之陵（位于印度，胡马雍为莫卧儿帝国皇帝） U2B

Tower of Babel 巴别之塔（圣经中所传的巴比伦通天塔） U8A

Turk 突厥，土耳其人 U8C

U

U.S. Steel Tower 美国钢铁大厦（芝加哥） U7C

Unity Chapel 唯一教堂 U4A

Ur 乌尔城（古代美索不达米亚南部苏美尔的重要城市） U9A

Urubamba River 乌鲁班巴河（秘鲁境内的河，发源于安第斯山脉） U8B

Usonian house "桑年"房屋（赖特自创的具有美国本土特色的房屋） U4A

Ustad 'Isa 乌斯塔德·伊萨（泰姬陵的主要建筑设计师） U8B

V

Vatican City 梵蒂冈（罗马教廷所在地，在意大利罗马城西北角的梵蒂冈高地上） U2B

Velázquez 韦拉兹奎兹（1599–1660，西班牙画家，画风写实） U3A

Vespasian 韦斯帕西恩（古罗马皇帝，开始营建古罗马圆形大竞技场） U8B

Vicenza 维琴察（意大利东北的一座城市） U5A

Victorian architecture 维多利亚式建筑 U1A

Vietnam 越南 U6A

Virginia 弗吉尼亚州 U3B

W

Wagner 瓦格纳（全名理查德·瓦格纳，德国作曲家，尤以其浪漫歌剧著名，常以德国的传说为其作品基础） U8C

Walter Gropius 沃尔特·格罗皮厄斯（1883–1969，德裔美国著名建筑师，首创金属构架玻璃悬强建筑） U2B

Warring States Period 战国时代 U6B

Webb & Knapp 韦伯·纳普（建筑公司） U4B

Wedgwood 韦奇伍德装饰陶瓷（商标名称） U3B

Weiyanggong 未央宫 U6B

Wenyuange 文渊阁 U6B

William Randolph Hearst 威廉·伦道夫·赫斯特（美国报刊和杂志出版商） U9B

William Strickland 威廉·斯催克兰 U3B

William Zeckendorf 威廉·柴根道夫（美国房地产开发商） U4B
Wiltshire 威尔特郡（英格兰南部） U10A
World Fair 世界博览会 U8C
World Heritage Organization 世界遗产组织（联合国教科文组织下属机构） U8B

Y

Yemenguan 雁门关 U9B
York Minster 约克大教堂（英国） U2A
Yucatan 尤卡坦半岛（中美洲北部） U8B
Yungang Grottoes 云岗石窟 U6C

Z

Zeus 宙斯（希腊神话中的主神、万神之王，天界的统治者） U2A
Zhou Dynasty 周朝 U6A
Ziggurat 金子形神塔 U9A

Appendix III: Recommended Books and Websites

Books:

Words and Buildings: A Vocabulary of Modern Architecture (Adrian Forty, Thames &Hudson Ltd., 2000)

Sir Banister Fletcher's A History of Architecture (20th edition) (Dan Cruickshank, Architectural Press, 2001) ([英] 丹·克鲁克香格,《弗莱彻建筑史》(第20版) 知识产权出版社/中国水利水电出版社, 2001年9月)

Encyclopedia of Architecture (Doreen Yarwood, Facts on File Publication, 1986)

Timeless Architecture (Elizabeth Meredith Dowling, Schiffer Publishing. Ltd., 2004)

Architecture: Design, Engineering, Drawing (6th edition) (William P. Spence, Glencoe/McGraw-Hill, 1991)

Frank Lloyd Wright: His Life and His Architecture (Robert C. Twombly, A Willey-Interscience Publication, 1979)

The Logic of Architecture: Design, Computation and Cognition (William J. Mitchell, The MIT Press 6th printing 1998, copyright 1990)

A Guide to the World's Greatest Buildings (Trevor Howells , Fog city Press, 2005)《建筑类专业英语》(李明章, 中国建筑工业出版社, 1997年6月)

Ancient Chinese Architecture (楼庆西, 外文出版社, 2002年1月)

Software:

Microsoft Bookshelf 2000 (微软百科书架2000版)

Websites:

http://architecture.about.com

http://encarta.msn.com/encyclopedia

http://www.new7wonders.com

http://education.yahoo.com/reference/encyclopedia/entry/architec

http://www.sanford-artedventures.com

http://www.architecture.com/go/Architecture/Careers/Jobs_4990.html

http://www.pritzkerprize.com